International Relations Theory and Ecological Thought

Towards a synthesis

Eric Laferrière and
Peter J. Stoett

London and New York

First published 1999
by Routledge
11 New Fetter Lane, London EC4P 4EE

Simultaneously published in the USA and Canada
by Routledge
29 West 35th Street, New York, NY 10001

© 1999 Eric Laferrière and Peter J. Stoett

Typeset in Baskerville by
BC Typesetting, Bristol
Printed and bound in Great Britain by
Creative Print and Design (Wales), Ebbw Vale

All rights reserved. No part of this book may be reprinted or
reproduced or utilised in any form or by any electronic,
mechanical, or other means, now known or hereafter
invented, including photocopying and recording, or in any
information storage or retrieval system, without permission in
writing from the publishers.

British Library Cataloguing in Publication Data
A catalogue record for this book is available from the
British Library

Library of Congress Cataloging in Publication Data
Laferrière, Eric
    International relations theory and ecological thought: towards a
synthesis/Eric Laferrière and Peter J. Stoett.
        p.   cm. – (Environmental politics)
    Includes bibliographical references and index.
    1. International relations–Philosophy. 2. Environmental
protection–International cooperation.   I. Stoett, Peter J. (Peter
John). II. Title. III. Series.
JZ1324.L34 1999
363.7–dc21   98-44174

ISBN 0–415–16478–8 (hbk)
ISBN 0–415–16479–6 (pbk)

Contents

Series editors' preface viii
Preface xi

1 Introduction: unearthing theoretical convergence 1

Introduction 1
Divisions of IR theory 5
Divisions of ecological thought 13
An ecological reading of IR theory: some guidelines 17
Conclusion 19

2 Ecological thought: a synopsis 22

Introduction 22
Utilitarian ecology 26
   Philosophical roots 27
   Utilitarian anti-environmentalism 30
   Conservationism 32
   Assessment 40
Authoritarian ecology 42
   Ecology and the Green Leviathan 42
   Ecofascism 46
   Gaia and misanthropic ecology 47
   Assessment 49
Radical ecology 51
   Selected sources of radical ecological thought 53
   Deep ecology 60

	Social ecology	63
	The status of ecosocialism	67
	Ecofeminism	69
	Assessment	71
	Conclusion	72

3 Realism and ecology — 75

Introduction — 75
Evolution of the realist worldview — 77
Addressing realist tenets — 83
   The realist ontology of conflict — 87
   Realism and hierarchy — 91
   Realism and homogeneity — 93
   Materialism and immutability — 95
   Realism as reductionist epistemology — 98
   An ecological assessment of realist tenets — 100
Conclusion — 104

4 Liberal IR theory and ecology — 106

Introduction — 106
The evolution of liberal IR theory — 108
Addressing key liberal tenets — 114
   Universalism in liberal IR theory: roots — 115
   Liberal universalism: the contemporary literature — 121
   The utilitarian basis of order — 126
   State and technocracy in liberal thought — 128
   An ecological assessment of liberal tenets — 133
Conclusion — 136

5 Critical IR theory and ecology — 138

Introduction — 138
IR theory and global process — 140
The normative critique of IR theory — 147
An ecological assessment of critical IR theory — 151
Radical ecology as critical IR theory? — 156
Conclusion — 162

| 6 | **Conclusion: maintaining a reluctant dialogue** | 164 |

Introduction 164
Summary of the book 165
Ecology and mainstream IR: assimilation or two
   solitudes? 168
Conclusion: some thoughts on future research 172

Notes 175
Bibliography 183
Index 205

Series editors' preface

While concern for the human habitat and ideologies affirming the vital link between mother earth and the humans species have had a long history and a constant appeal, the past three decades have witnessed a surge in an awareness that humanity is inflicting on itself permanent and possibly irretrievable environmental damage. This series of books on the politics of the environment aims to provide the information and the perspective needed for an understanding of this predicament, of the anxieties to which it has given rise, and of the steps that are being taken at national and international level to address the problems that it poses.

The urgency of the environment predicament has already produced a substantial corpus of publications, and that corpus is constantly growing. The present series covers three broad areas. The first consists of the ideas and debates that the environmental movement has generated. There is room in the series for treatments of both speculative and practical contributions to those debates, the aim being to engage in analysis rather than advocacy. Second, the series contains analyses of the fortunes of the various political movements and organisations that have environmental goals. These range from inchoate and spontaneous collective action to the more organised and abiding political parties and non-governmental organisations. At the same time, the environmental policies that other political parties have been led to adopt are included, even in cases where those parties espouse ideological positions distant from those characteristic of green parties and movements. A third concern of the series is policy-making processes at national and international levels and, increasingly, the processes of trying to implement programmes to tackle existing environmental degradation in ways that do not simply worsen the problems and create new ones.

The emphasis, at least in the preliminary stages of the series, is on the advanced industrial countries. However, the series editors are fully aware of the interconnectedness of environmental issues and of the essentially international nature of environmental threats and of attempts to address them. While, therefore, it is not possible to include coverage of the environmental problems of individual developing countries, treatments of the broader international issues will be welcomed and, in particular, those that highlight debates and developments concerning the global North–South issue.

The result is a comprehensive but manageable focus on the politics of the environment such as the series editors believe is needed as the twentieth century turns into the twenty-first century. As for the expected readership, the books in the series will carry the original research, but of an accessible kind. Many of the books will therefore have the character of a 'textbook plus'.

In conclusion, no scholarly endeavour should shy away from conveying a message of some sort if it is to catch imaginations and monitor the impact of change, as this series aims to do. While it is not the intention of the series editors to make any overt political statement, we are prepared to express a concern, based upon what we believe to be incontrovertible facts, from which everyone alive today must be prepared to draw the conclusions. Some of the resources on which we depend are finite and irreplaceable; the world we shall leave behind us will be, in many respects, worse than that which we inherited, and almost significantly worse; and the processes of discerning the signals of danger – framing responses to them, achieving agreement on the action to be taken to counter the danger and actually taking that action – are slow and difficult. Only one step of imagination is demanded of the reader beyond measuring these facts. This is to make the assumption that achieving the tasks imposed by the present environmental predicament is possible. Unfortunately, nothing at present could be less evident. What is evident is that we can escape the consequences of failing to cope with those tasks by returning to the earth from which we came, but that no-one escapes the responsibility for failure.

Michael Waller and Stephen Young

Preface

As doctoral students in political science in the early 1990s, we witnessed a burgeoning academic interest in global environmental politics in the wake of the celebrated Brundtland Report. Many students of comparative politics, political philosophy, and ecology had shared a common effort: trying to grapple with the impact of environmental problems on societal relations within the industrialized and southern states. Though various representations of the ecological crisis had surfaced much earlier (most conspicuously, the debate on overpopulation and energy supplies in the 1970s), and non-governmental organizations had popularized the plight of rapidly declining whale populations and the intrusive character of industrialization, social scientific and philosophical studies in this case took a few years to catch up.

In the field of international relations (IR) theory, however, work was embryonic even at the turn of the last decade. Policy-oriented analyses of environmental diplomacy had appeared, yet few scholars explored its metatheoretical implications. Instead, we learned about regime formation, an old concept by that time, in a new and exciting area where policy convergence among states was becoming increasingly necessary. While IR theory has always been visited by, and has borrowed liberally from, various strands of political philosophy, the searching and challenging body of ecological thought which had emerged during the preceding decades (and, indeed, centuries) had yet to have a substantive impact. When IR theorists did discuss ecological issues, they did so from a very conservative understanding of the ecological problematique, reformulating classic problems of scarcity and collective action. This is still largely the case.

As we analyze the perspectives shaping IR theory, we can presumably identify several strands of ecological thought built into those perspectives and thus broaden the links between the two disciplines. Surely, the scholarly community is well past the initial assertion that environmental issues are transboundary, and thus within the domain of IR. Similarly, we should go beyond the commonly accepted point that security requires a redefinition extending to environmental threats. Still, we should not minimize the extent to which those early studies, accounting for problems such as global warming, species loss and deforestation, have served to demonstrate the integrated character of politics: the inseparability of governance, economics, citizen participation, development strategies, international organizations, and many other facets of global politics itself.

This integrative objective very much motivated our thinking and writing of this book, and thus our particular interest in synthesizing political ecology and IR theory. Is there an identifiable commonality, given the importance of the environmental problems upon which the scholarly community has, finally, been forced to reflect? Both political ecology and IR theory have philosophical roots, and so we wanted to dig in this conceptual garden, see where roots had become entwined – or ensnared? – and thus explore potential paths of new growth. This may be especially fruitful in the case of the more radical strains of ecological thought, which have little, if any, representation in the mainstream of IR theory.

We proceeded by establishing our own parameters of recent ecological thought, with an eye on its philosophical sources. The dominant western perspective has, without doubt, been the utilitarian one, which sees nature as valuable insofar as it is useful for human/societal development, and leads at most to conservationism. But other strands have emerged with vigor in the last century: what we call authoritarian ecology, and several variations of radical ecology, including deep ecology, ecosocialism, ecofeminism, and social ecology. Next we examined three central schools of IR theory: realism, liberalism, and critical theory. We have attempted to identify both the ecological tenets inherent in these approaches as well as the potential for contributing to the further development of a synthesis between ecological thought and IR theory. What emerges is an ongoing dialogue between two fields of thought which have more in common than either may suspect, and yet which remain virtual strangers, despite the fact that they are both struggling with many of the same problems.

We conclude that radical ecology, which in various forms challenges the mainstream of both western society and its intellectual apparatus, may be best viewed as a critical perspective within IR theory itself. However, it needs to take explicit consideration of international questions into account, and at present there is great room for such conceptual development. In other words, it is not just the study of ecology that can enrich our perspective on IR theory; ecologists would do well to familiarize themselves with IR as well. We hope this book offers an exploratory beginning toward this end.

Our scholarly association began in February 1992 at a pre-United Nations Conference on the Environment and Development (UNCED) symposium in Victoria, British Columbia. Several years later, International Relations Theory and Ecological Thought took shape as a response to Routledge's announcement of a forthcoming series on global environmental issues, and we are grateful to one of its editors, Stephen Young, for encouragement in both our early careers. Eric had recently completed a doctoral dissertation at McGill University analyzing the concept of peace in IR theory from an eco-radical perspective, and that dissertation provided the intellectual inspiration for the broader project realized here. Peter's work has been more policy-oriented, but the "theory bug" bites hard and fast amongst academics, and he fulfills its demands with this book.

As such, this book is a synthesis of styles and orientations. This can make for trying writing, but mutual fascination with the subject matter, frustration with the exclusion of ecological thought in the realm of conventional IR theory, and friendship overcame the many sticking points. A spirit of collaboration and mutual criticism prevailed, with much of the editing accomplished over long-distance telephone in the cheaply rated wee hours of the night. But there are several people who looked at earlier drafts of chapters to whom we are indebted.

The first and foremost of these is without doubt Pat Romano, Eric's insightful and patient wife. She was always prepared to offer supportive and critical comments, even in the midst of those three-hour telephone talks. We thank also Robert Boardman, Bill Graf, Eric Helleiner, Doris Miller, Shane Mulligan, Jorge Nef, and Matthew Paterson for comments on earlier drafts. We are grateful to the editorial and promotions staff at Routledge, and Peter thanks the Social Sciences and Humanities Research Council of Canada for financial support. Naturally, we alone are responsible for all errors and shortcomings.

Finally, we dedicate this work to Pat, Ryan, Alexandra, and the memory of Stella Mazlum (1908–1998) and Berc Mazlum (1907–1998).

Montreal and Ste-Anne-de-Bellevue, September 1998

Chapter 1

Introduction
Unearthing theoretical convergence

INTRODUCTION

Where are we, several years after the widely celebrated, then roundly denounced, Earth Summit of 1992? Occasional media coverage of the odd hurricane or typhoon aside, has the environment died as a popular issue? It rarely constitutes a major electoral issue any more, if it ever did. The haphazard implementation and, often, dismantling, of environmental policy continues in Northern and Southern states. Thousands of bureaucrats now work with such policies on a daily basis, but it is within the confines of a shrinking public sector. The academic community that was born and raised on the wings of environmental concern has matured, but still grapples with fundamental divisions in the definition of seemingly ubiquitous terms, such as the most popular but perhaps most meaningless, *sustainable development* (see for example Fisher [ed.] 1995).

Within the field of international relations, the environment is no longer an oddity, or a tack-on; it has become one of the standard and illustrative issue-areas, referred to at introductory lectures and graduate seminars alike. But in this field, like any other, conceptual and epistemological divisions persist. Just how important environmental problems are – to states, or state actors, in particular – remains a matter of considerable dispute. More fundamentally, has the need to acknowledge the importance of this issue-area changed the theoretical apparatus utilized by scholars of global politics? One might argue this occurred, to some degree and within some strands of the discipline, when nuclear weapons became the dominant concern during the Cold War; or when the evolving North–South conflict pushed theories of imperialism back into the classroom. The intellectual journey leading to this book began when we asked whether or not such a shift had occurred because of the environmental crisis.

At the same time, environmental science has evolved at an accelerated rate as business goes green and new technical subfields open to entrepreneurial innovation. *Agenda 21*, if ever actually implemented in any manner approaching its entirety, would create a virtual torrent of new jobs, investment opportunities, and research projects.[1] But what of the softer science of the environment, our conceptual understanding of its, and our, place in the greater scheme of things? What about the understanding of ecology in its social context, or ecophilosophy? What about environmental ethics? Here, as in international relations theory, we see old splits, largely ontological, that have solidified over the years. We might note also a certain reclusiveness: ecologists risk caricaturization when they sound like alarmists all the time; their subtler but important social thought is often obscured by the need to present "the crisis." This has been true in general, but especially when attempts are made to link ecological social theory with politics, and global politics in particular. Thus, the second path our journey encountered is the one asking what effect the need to "think globally" has had on ecophilosophy.

The purpose of this book is to bridge the gap between two fields of social theory: international relations theory and ecological thought. Rather than the miraculous production of a complete synthesis, however, we seek an opportune cross-fertilization between the two. The formal study of global politics has acknowledged the environment, but little IR theory explicitly incorporates ecological principles, despite earlier work that sought to utilize "environment" in variants of the systems approach (Kaplan 1957; Sprout and Sprout 1971), and despite the present wealth of policy-oriented work that deals directly with resource regime questions (Haas, Keohane and Levy [eds] 1993; Spector, Slostedt and Zartman [eds] 1994). And, if it is possible, ecologists have been even less willing or able to incorporate the vital questions raised by the condition of international politics (or, more precisely, the important debate about what that condition actually is) in their work. Much of the ecological literature produced in this century reads as though it were written with a political system roughly the size of Thoreau's Walden Pond in mind, though it is clear that our present concerns are indeed global in scope.

We proceed on this conceptual journey with certain assumptions, explicated immediately below:

1) The physical health of the planet is indeed in danger. V all analysts are prepared to acknowledge a full-fledged e_ crisis, few would dismiss the various warning signs emanating from nature: the warming of global climate, thinning of the stratospheric ozone shield, accelerating loss of biodiversity, acidification of fresh water lakes, soil erosion, desertification, and many others. The immediate causes of such alterations are well-known, from unsustainable consumption patterns, widespread releases of toxic material, high populations in some areas, and inflated standards of living in others. All in all, planet Earth's "biotic capacity" has been put into serious question. This is neither to suggest that life patterns on Earth have been immutable nor to deny the fact that human societies have wreaked havoc with their environment in centuries past. Rather, it is to recognize the formidable scale and speed at which ecological degradation has operated in the late contemporary period.[2] The environmental crisis is, in essence, the most profound contribution to what Ernst Haas, in his discussion of "eco-reformers," refers to as *the problématique*.[3] Despite the obvious signs of continued decay, however, we add a significant, if hopeful, caveat to this assumption: *it is not too late*. The most harmful anthropogenic causes of environmental harm can be reversed; nature has tremendous self-healing powers.

2) In most cases, the underlying causes of ecological degradation are political; and where they are not, the human consequences of natural disasters, including maldistributed relief, are. The "attack on nature" is symptomatic of a commitment to material growth and state power, which requires the systematic control and use of human and non-human nature. The roots of this social project can be traced back to the momentous intellectual and political developments of the seventeenth century, where Newtonian science and the nation-state arose as twin pillars of modernity. With the development of capitalism, contract theories effectively abolished the organic character of communities, leaving the presumably self-interested individual to survive in a competitive world. Utilitarian theories of the late eighteenth and early nineteenth centuries were logical consequences of the materialistic and mechanical "redesigning" of the world. Contemporary ideologies of growth and power – and, in popular parlance, prevailing conceptions of sustainable development – have harnessed the forces of science and technology to create large markets for high value-added goods. In the process, "natural resources" have been mined at staggering levels,

with often disastrous ecological consequences and violations of basic human rights. In general, the state has continuously perpetrated the ideology of control necessary for this process, though it has also acted to mitigate some of the environmental excesses that result from it. There have been substantial intellectual challenges to the hegemony[4] of this ideology as well, though we should avoid the error of categorizing all forms of environmentalist thought as anti- or post-modern.

3) If the causes of ecological deterioration are political, and assuming we need not give up hope on the prospect that such deterioration can be reversed, so need be the solutions. This is evident in both reformist and radical senses: whether we are enacting energy taxes, passing wildlife protection legislation, negotiating treaties to establish regulatory regimes for marine oil pollution, or resisting the communal value of corporate culture and centralized decision-making, we are engaging in politics. Ultimately, we believe the rehabilitation of nature entails some form of commitment to an "ecological society," which itself depends on some conception of economic and political order. However, such significant changes are difficult to achieve where institutions are firmly entrenched and enable these social events. In the international system, as the *realist* school introduced later in this chapter constantly reminds us, the prevailing setting is not conducive to such changes, though the dynamics may be different at the state or sub-state level – opening a wide range of questions about the role of the state, non-state actors, international institutions, and other actors in the process.

4) Following from the above, the nexus between ecological thought and IR theory is most apparent when both are understood as *political* theory. The historical "mission" of IR theory is both analytical and prescriptive, seeking to understand the conditions for the maintenance and breakdown of order among large sovereign entities, usually states. Yet this theoretical base has exploded in recent decades as IR scholars struggle to deal with new-found problems, within and then out of the Cold War context. Meanwhile, ecological thought seeks to uncover the political sources of ecological deterioration and explores political structures for ecological living – tasks that logically carry into the international/global realm. We assume that the nexus between these two fields of political theory is relatively underexplored, and that both will benefit from a sustained dialogue. Further, we assume that, since modern social theory is largely premised on a body of philosophical work, we are under no

methodological obligation to draw a sharp distinction between theory and philosophy. Finally, it is evident to us that theory construction is both a descriptive and prescriptive exercise; though their identification can only be a matter of interpretation, all theories are premised upon normative assumptions.

In sum, it may be argued that the generalized attack on nature in the late twentieth century compels an interdisciplinary rapprochement between IR theory and ecological thought (or "ecopolitical" theory). For the latter, its effort at political design should benefit from the studies of IR theorists into the realities of power and the possibilities of cooperation among large actors (states, classes, communities, societies). Yet the stakes are probably higher for IR theory, since much of ecological thought is already well informed with political theory. In other words, by looking at environmental problems as political problems, ecological thought is arguably well positioned to help IR theory refine its own understanding of order, peace, security, and power (its traditional explananda). In fact, ecological thought can play a role in reshaping both the normative commitments of IR theory (which are rarely articulated explicitly) and its conceptions of political process (Laferrière 1996).

In this introduction, we take the initial steps in formalizing a dialogue between IR theory and ecological thought. We first define the contours of IR theory, offering some preliminary comments on the trichotomy adopted in the book and detailed in Chapters 3, 4 and 5. We then briefly introduce another trichotomy, organizing the numerous divisions within ecopolitical thought and upon which the lengthy discussion in Chapter 2 is based. Finally, before closing with concluding comments, we provide a framework, based obviously on ecological thought, for the particular treatment of IR theory found in Chapters 3, 4 and 5.

DIVISIONS OF IR THEORY

The diversity of theoretical literature in the study of IR necessarily complicates any neat effort at categorization. IR is a highly diverse field of scholarship, ranging from purely theoretical reflections to straightforward policy-oriented research. Many IR scholars have charted "paradigmatic maps" of the field, yielding standard three- or four-fold divisions; "sub-paradigmatic" break-downs could, of course, produce even more diversified statements of perspectives and approaches, many of them reduced to the work of a handful of scholars.

For our purpose, we feel comfortable in retaining a simple structure that would capture the main ontologies and epistemologies of the field. Ultimately, we want to know why IR has addressed ecology in only limited terms, and the answer to this question does not require an account of all theoretical nuances. Therefore, we acknowledge Robert Cox's celebrated distinction between problem-solving and critical literature in IR, yet we also accept another distinction, that between realist, liberal, and critical schools (nonetheless recognizing their several commonalities).[5] Even more broadly, perhaps, we can distinguish between the "mainstream" of IR theory, or the more prevalent approaches – namely, realism and liberalism – in the traditional halls of academia, and the critical or radical school.

It would be premature to declare a winner of the battle for sole possession of the title "mainstream" in IR theory. For decades, scholars have assumed and taught that realism, in its classic form if less so in its latter day "structural" appearances, has been the dominant perspective (indeed, some would employ the more demanding term, paradigm) in the discipline. This was largely a consequence of American prevalence in the academe, and more precisely the early domination of the formative and forbidding text by Hans Morgenthau, *Politics Among Nations* (1948, 1993). Prior to this publication, former British diplomat Edward Hallet Carr's classic *The Twenty Years' Crisis: 1919–1939* (1946) had presented a rough dichotomy between an idealist and realist view of world affairs; but it was Morgenthau's more rigid and comprehensive work that provided generations of American students of IR with their main textbook. Other prominent texts that were less widely used at the time read similarly: they are written from an American vantage point and informed by the realist perspective on the twinned concepts of security and national interest (e.g. Hartmann 1957).

Disciplinary convention paints a linear picture of the evolution of IR theory: classic realism dominated, and, aided by a trenchant, perhaps even pre-emptive critique in Kenneth Waltz's *Man, the State, and War* (1959), survived a challenge by behaviorialism; then theories that forced the issue of interdependence onto the agenda rose and have continued to struggle with realism for overall acceptance. Meanwhile realism has split into the classic, which emphasizes the causal role of human nature, and the systemic or structural type, again brought forth most popularly by Waltz; and the interdependence schools have branched into institutionalist and reflectivist variants. Radical theory, meanwhile, has always had a small if devoted following in the United States, Europe, and elsewhere. We should note, however, that this is a largely American reconstruction of the path of the discipline. The British tradition focuses much more clearly upon the development of an international society (e.g. Bull 1977). Meanwhile, the liberal school, which stresses the cooperative nature of political agents given their own self-interest, has gained considerable popularity within the discipline, particularly in North America and Europe. A concern with the formation of international institutions has led to an immense volume of related literature that, one might well argue, has a liberal premise.

Morgenthau's classic contribution was the explication of his six principles of political realism. Politics is governed by objective laws, which can be held as operative assumptions found in human nature, and we can test this by asking what rational choices decision-makers face, where their preferences will lie, and then observing their behavior. Interest is defined as power for the purpose of political science. Though the pursuit of interest is constant in world history, types of interest vary according to context. The state is moved by the moral principle of national survival, which requires prudence (reflected by cost/benefit analysis). There is no knowable good and evil as pertains to state interests (a fundamental amorality reigns in world affairs). And, finally, political realism is based upon a pluralistic conception of human nature; it argues that people are not merely political animals, but that we can nonetheless separate the political from the religious, economic, or moral person.

Just as Morgenthau's textbook defined IR theory for several decades, the present conception of realism is largely derived from a popular American text, namely Robert Keohane and Joseph Nye's *Power and Interdependence* (1977). That is, realism is defined as not being complex interdependence; it is limited to concerns over state power and military power especially, and only (unified) state actors matter in world politics. The origins of realism, despite Machiavelli's usage of the term in a secular plea for power-oriented rationality, are philosophical; they are, even, moral in tone and urge. The roots of realism can be traced back to such immortal scholars as Kautilya, Thucydides, Machiavelli and Hobbes, often referred to in the classroom as the "great quartet." However, there will be no attempt here to provide yet another exegesis of the *Arthashastra*, the *History of the Peloponnesian War*, the *Prince*, or the *Leviathan*.

In IR theory, realism rests upon the concept of anarchy, defined in terms of the lack of an authoritative, unifying government amongst nation-states. This condition of anarchy amongst self-interested, essentially autonomous states is the key to a systemic understanding of realist orientations. Put bluntly, "there is a constant possibility of war in a world in which there are two or more states seeking to promote a set of interests and having no agency above them upon which they can rely for protection" (Waltz 1959: 227). The ontological orientation of realism is one of conflict and aggression, since it entails the basic, and immutable, existence of national hierarchies and this compels the conceptualization of an atomized world of colliding centers of power. The underlying philosophical assumption is the universality of the desire for power, held to be common to all men.[6]

This desire emerges once survival has been secured, and ensures the *permanence* of conflict throughout human history and into the future. Thus the ontology of realist politics is biopsychologically rooted; the desire for power becomes sufficiently institutionalized through states that it becomes a national duty to pursue it in the international arena. Raymond Aron, for example, sees the state system as a state of nature, in which conflict and aggression predominate: "the necessity of national egoism derives logically from . . . the state of nature which rules among states" (Aron 1966: 580).[7]

As a whole, then, realists believe that power drives are natural, even rational, that political associations (states or similar finite entities with hierarchical organization) are natural, and that the former are served by the latter. In this natural state, the strong pursues the weak, the weak fears the strong, and both use physical resources to survive or fulfil their natural mission. In fact, survival also animates the strong, who know not only that their life essence is in fighting, but that the weak may grow to be strong too. Fighting may be delayed by the achievement of a balance of power, where mutual deterrence becomes a temporary strategy; but this is merely a delay as the cycle of war and peaceful preparation for war continues.[8] Nature is cruel: as the German historian Heinrich von Treitschke would write, "the features of history are virile, unsuited to sentimental or feminine natures . . . the weak and cowardly perish, and perish justly" (Treitschke 1963 [1916]: 31).[9] Survival is possible through strength, and in international politics this translates most readily into the need for military power.

Classical realism is (in)famous for its pessimistic view of human nature, as well as the still widespread tendency to view political actors as mechanically colliding bodies. In other words, realism presumes an analysis of the human psyche that stresses both self-interest and a quest for power, ensuring a permanent state of war between sovereign entities; at the same time, realism includes a mechanical ("structural") model which attributes motives based on actors' positions within a political system. This assumption base (self-interest, competition, and mechanicism) is, admittedly, not limited to realism, though realism does complement it with an unrivalled emphasis upon the causal power of anarchy in world affairs. However, there is an important difference between the circular pessimism of the "power politics school" and the linear liberal optimism regarding the "good life;" as we will see, this distinction has ecological implications as well.

If realism may be rooted in the "amoral" policy recommendations of Machiavelli to his Prince as well as in Hobbes' mechanical worldview, *liberalism*, as the conceptual banner of modernist thought, cannot be conceived as a complete alternative to realism. After all, Hobbes may well be depicted as a pre-liberal, while the fundamental separation between ethics and politics demanded by Machiavelli does characterize much of the contemporary liberal literature. Realists and liberals share a contractualist understanding of society, and both have endorsed the basic separation between fact and value, subject and object that is characteristic of positivist social science. In sum, realists and liberals constitute the problem-solving mainstream of IR, and it was in fact an attempt to marry the two, or the innovation of Robert Keohane and Joseph Nye's *complex interdependence*, that generated much of the regime-based ecopolitics literature.

Liberal IR theory is most clearly distinguished from realism by its relatively optimistic assessment of human nature (or, conversely, in some cases, the refutation of its significance) and its policy prescription for global freedom obtained through institution-building. While realists relinquish freedom and prudently search for social stability, even if it implies an armed stability, liberals are to varying degrees committed to sustained social progress – freeing the individual from the constraints of traditional hierarchies and from nature itself. In a book on ecology, this is reason enough to divide the mainstream for purposes of review.

Basic liberal assumptions include the notions of incremental, if not inevitable, progress; unity based on common experience and incentives; and the sheer value of reason. Liberal thought has a close connection to thinking about domestic, or intrastate, politics. For example, R.D. McKinlay and R. Little (1986) divide liberalism into "pure" and "compensatory" streams, corresponding roughly to popular images of the conservative right (or, as Canadians have often referred to it, the business liberal agenda) and proponents of the welfare state. The individual, rational decision-maker remains the core of the liberal ontology, however.

More than any other widespread branch of political theory, liberalism carries the banner of modernity. Zacher and Matthew (1995: 110), in their comprehensive overview of the field, argue an important thesis of modern liberalism in IR theory is that international relations are "being transformed by a process of modernization that was unleashed by the scientific revolution and reinforced by the intellectual revolution of liberalism." This is not to be confused with the modernization paradigm of development studies so widely discredited in decades past, though the confusion might be understandable. This modernization process has five core components: liberal democracy or republican government; international commercial and military interdependence; shared cognitive progress; international sociological integration; and international institutions.

Liberalism subscribes to the idea that peace can be pursued simultaneously through several avenues. This includes cosmopolitanism, which echoes the writings of Immanuel Kant, emphasizing a gradual cultural convergence of individuals and nations; this resonates particularly loudly in the age of supposed "globalization."[10] Kant's writings have become common reference for liberalism in IR theory, though some controversy surrounds this practice.[11] And of course the great stimulant of cosmopolitanism is trade, as open as possible within a world of sovereign states.[12]

Part of the liberal project, then, is to build bridges between nations and cultures, so as to realize material security (through global comparative advantage) and the wider ideal of human unity. Open commercial lanes will increase the chances for peace by increasing material bounty, directly reinforcing the (political) rapports of friendship. Enmity is dissuaded by the inevitable increasing of its "opportunity cost;" war becomes prohibitively expensive in an interdependent world economy, especially when cross-investment prevails. This not only facilitates further liberal developments, but injects an element of stability, manifested in contractual relations between states, into the global system.

Modern liberal thought in IR has of course become much more nuanced than this rough caricature. However, the essential message is that of recognizable contract zones, accommodation (not necessarily the more ambitious harmonization) of interests, and organizational learning through time. From the early neo-functionalist work evolved a flurry of theoretical variants, including Peter Haas' (1990) focus on "epistemic communities," Oran Young's (1989a) analysis of "institutional bargaining," and Ernst Haas' (1990) work on organizational learning. There is another avenue to peace stressed by liberal thought, and it is that of peace-through-technocracy: this centers on the depoliticization of problem-solving dynamics in an essential context of mutuality of interests. This, coupled with the imperative of expanding trade and trade choices, relies on a further possibility, that of peace-through-rules; more specifically, the institutions that can create rules and foster contractual environments for further rule-making (i.e. flexible yet meaningful institutions); less optimistically, perhaps, institutions will enmesh political actors within routinized patterns of cooperative behavior. Twentieth-century internationalism may be seen in this light, a modern celebration of the Grotian ideal, fully endorsed by contemporary liberal theorists of IR as a background for process-oriented arguments derived from social-choice theory.

Standing in opposition to the "problem-solving" approaches of realism and liberalism is a *critical* literature, whose actual range is still a matter of debate amongst both adherents and observers. Conceived broadly, it surely includes the neo-Marxist literature of the dependency and world-system theorists; the latter may not have engaged in a direct dialogue with established scholars of IR, but their problematique is undeniably that of political relations between large "sovereign" actors (state and class). More visibly, however, strands of the subfield of international political economy are often considered to be rooted in Marx, if not because of an explicit commitment to class analysis, then for its emphasis on structural power and normative concern with social injustice. If anything distinguishes the critical school from the mainstream, however, it is the rejection of the state as the solution to modern problems. Realists see the state as a problem, in conditions of anarchy, but posit simultaneously that the solution to the problem of survival is a strong state, or at least a balance of power amongst states. Liberals see the individual as the ultimate sovereign, to be sure, but see the state as a primary institution that can advance the human condition. This is increasingly challenged, however, in the age of the ideology of *globalization*: many neoliberals would argue the state has less relevance in this grand project today.

Critical theorists, whether or not they subscribe to an orthodox interpretation of the state as a vehicle for class domination, tend to see the state as a representative of entrenched interests, and an obstacle to achieving a less conflict-prone global society. In addition, there is an overt concern with social justice, as it pertains to gender, legal, cultural, and environmental issues. In particular, the sacredness of the institution of sovereignty receives direct challenges. As the liberal literature would also insist, it is gradually eroding, but critical theorists have argued it also perpetuates a social structure inimical to world peace (Walker 1990).

The critical literature could also incorporate the normative reflections of the World Order Models Project (WOMP) scholars, many of whom directly addressed (yet were largely ignored by) mainstream IR theory. The WOMP literature's attempt at reinstating normative theory within the field is worthy of the "critical" epithet; yet it should also be noted that significant strands of that literature, in their policy recommendations, squarely follow a model quite similar to that espoused by the liberal "problem-solvers" above. In short, the WOMP is part of a broader movement at reforming both IR thinking and international institutions. While it questions the positivist epistemology of IR theory, its policy perspective is not necessarily innovative. However, this is not to deny the WOMP's obvious understanding of the various problems affecting global order, which is, in fact, sympathetic to both neo-Marxist and reform-liberal analyses (Falk and Kim 1982).

Alternatively, critical IR theory may be limited to more recent attempts at questioning the epistemology and the social mission of IR theory. This school is inspired by both the Critical Theory of the Frankfurt School and French postmodernism. Here we see the evolution of a schism: both strands of critical theory engage in a deconstruction of the hegemony of IR theory and of foreign policy practice, yet only one engages in the "(re)constructive" project of modernity. In both cases, however, there is a normative commitment to emancipation, though this is a term that often lacks definition. This commitment is readily apparent in feminist critiques of IR (e.g. Grant and Newland [eds] 1991; Enloe 1989; Mies 1993), but no less so in the neo-Gramscian writings of Cox and others. As a whole, critical theorists seek to expose the limitations and, indeed, the normative underpinnings of positivist IR theory, while accounting for a broader understanding of political processes affecting global order – a predictable sequel to the WOMP literature. In essence, a critique of the predominant liberal spin on "globalization" is found (incidentally, a critique predating the popular consumption of the latter concept).

Now that we have presented a necessarily brief synopsis of the three central divisions of IR theory, let us turn to an even more cursory treatment of three equally representative divisions of eco-political thought. We will expand on all of these in subsequent chapters.

DIVISIONS OF ECOLOGICAL THOUGHT

Ecological thought has produced an extremely varied literature, much of which is derived from established philosophical traditions. Indeed, thinking about nature is an essential task of philosophy, an old enterprise dating back to such luminaries as Aristotle and Confucius. And of course all of the world religions have profound conceptions of nature and our relations with it (Jain [ed.] 1996; White 1967; Pepper 1989). Some conceptions are based on civilizational worldviews; for example, Ravi Ravindra (1991) writes of three basic views of nature: Western, Sineatic, and Indian.[13] We do not intend to explore these categorizations in significant depth, however.

The term ecology dates back only to the late nineteenth century, when the scientific community began to understand the relationship between the component "parts" of "ecosystems." Thinking on ecology was clearly spurred by the rapidly growing excesses of industrialization, as human intervention was now exposing the fragility of delicately balanced habitats. Ecology arose as a scientific discipline committed to the understanding and aspired "engineering" of ecosystems, but it developed simultaneously as a philosophical endeavor, searching for the larger (metaphysical) meaning of nature. Inexorably, ecology was to become an integral part of the twentieth-century critique of modernity, as awareness of an increasing number of environmental crises exposed the limits of capitalism, socialism, statism, representative democracy, and science itself.

We adopt an eclectic notion of ecology – as scientific field and as philosophical reflection, as a descriptive and ordering enterprise and as a normative investigation. Few fields make us aware of the extent to which science, and its applications, have both immediate and long-term social consequences. An initial discussion of "utilitarian approaches" should capture one distinct strand of ecological thought, the one which, incidentally, has been exclusively favored (or implicitly employed) by mainstream IR theorists. This version of ecology reflects what may be best termed a *managerial approach* to sustainability. We argue that this approach has various roots, from religious conceptions of stewardship to the writings of early naturalists in the United States and elsewhere. On a global level, the most popular expression of this perspective, which was combined with equal parts fanfare and liberalism when released in the 1980s, is the famous Brundtland Report, *Our Common Future* (WCED 1987).

Conservationism is an appropriate label for this brand of ecological thought: beyond sustaining our survival, nature may be legitimately conceived as a pool of natural resources to be used for the primary purpose of economic growth. As a managerial perspective, conservationism is at ease with technology and sustains the liberal dream of human freedom through control of nature; as a political philosophy, it remains unarticulated. The common utilitarian understanding of nature as "environment" has definite conservationist roots: environmental issues are in essence resource issues – scarcity issues, affecting human beings yet existing in a separate sphere of consciousness.

The next section of Chapter 2 will focus on what may be called "reactionary approaches" to ecology. Here we find attempts to tie nature and the rise of industrial society with a conception of community, building on a long lineage of naturalist arguments dating from Ancient Greece. Two distinct bodies of literature can be identified under this heading, mirroring to some extent the historical rift between rationalism and romanticism. On the one hand, several authors have achieved a remarkable degree of notoriety by advocating centralizing and/or survival-ethics solutions to ecological degradation. Their treatment of nature is largely utilitarian, and, having recognized the common-resource problem, their solution is to empower scientists and government officials (who are presumably politically neutral) and, in some cases, to save nature by excluding selected groups from the human "lifeboat" (Hardin 1974). The stability of "enlightened" hierarchy is hereby pursued, and expedient control/sacrifice of human beings is logically warranted so as to preserve our natural infrastructure.

Authoritarian policy recommendations in the wake of ecological crises form one identifiable, "reactionary" attempt at linking nature and politics; that literature is essentially rationalist, reacting to the problems posed by scarcity and overpopulation, two contestable but obviously interrelated terms. The label of *ecofascism* has been attributed to that literature, yet this term probably better designates the historical attempt at linking nature with an ideology of "blood-and-soil." Nineteenth-century romanticism would eventually produce a form of nihilism that evolved into the irrationalism of extreme-right movements, particularly Nazism, which relied on modern technology and pervasive military symbolism while paradoxically extolling the return to a simple, agrarian life (for the chosen, of course).

Nazism, however, was only one expression (however depraved) of a broader movement evoking a "back to nature" theme (Biehl and Staudenmaier 1995). For many naturalists of the past century, especially in the vast expanses of the New Continents, the rapid industrial take-over of nature constituted an absurdity, a perversion of the "biotic ideal" of humankind; thus arose a "preservationist" reaction, or the advocacy of future policies for the creation of national parks and the protection of endangered species. Such preservationism is not necessarily reactionary, by any means – in fact, much of it can coexist with conservationism. However, this romantic assertion of nature is also associated with a certain mysticism, which should not be confused with nazi "naturalism."

As expressed by some exponents of so-called "deep ecology" (see below), a conception of nature as the realm of the "extended self" may detract from a positive reassessment of political and economic conditions and from a reconstructive project towards human emancipation. Stated differently, a purely preservationist ecology may tend to belittle the value of humanness, blending it into an ecological whole stressing global balance. As we shall see, this may be the lesson to be derived from the Gaian, "unicellular" conception of planet Earth. This is not to suggest that the vision of Gaia is unhelpful or even unscientific (and it is certainly charismatic), but only to acknowledge the delicate ethical and political implications of such naturalism.

A final chapter in this section will concentrate on what we term *radical ecology*, whose treatment of "ecology" is geared towards a fundamental reappraisal of modern social structures for the purpose of emancipation from structures of domination. Radical ecologists are specifically social theorists, who see in degraded nature an attack on the weak and marginalized. In this sense, most (but not all) radical ecologists reject ecological anthropocentrism, which subordinates "nature" to "humanhood."

Our treatment will begin with ecosocialism, perhaps the only strand of radical ecology that still clings openly to an anthropocentric worldview. "Ecosocialism" may sound like a misnomer, linking Marxism (which fully endorses industrialism) with the preservation of nature (Eckersley 1988). However, there exists a distinct tradition of "red-green" scholarship and activism committed to both "nature" and the "common people," but granting ethical priority to the latter. In this view, the abolition of capitalism is a necessary condition for ecological balance, since capitalism, as a system, depends on the exploitation of all life forms. Quite interestingly, some *dependency* theorists have adopted ecosocialist principles when examining the North–South power relation, particularly as it involves the natural exploitation of the enclave economy.

In the work of Murray Bookchin, ecological thought and a refined version of anarchism join to produce a "social ecology" stressing the inextricable co-variance between social hierarchy and biotic loss/rarification. Ecoanarchism aims at rehabilitating the individual and broader nature, all at once, by decentralizing economic and political power; the ensuing ecological society is meant to revive the political egalitarianism of direct democracy within an eminently modern context (Bookchin 1989).

Bookchin's social ecology is usually understood in opposition to the approach of deep ecology, yet we will argue for a substantial convergence between the two. The attack on deep ecology usually rests on the relative lack of political theorizing among its proponents, in several cases translating into poorly conceived recommendations about the fate of human beings and, more generally, into a brand of mysticism which would deny a positive role for reason in an ecological society. In other words, deep ecology's communion with nature would appear reactionary, and perhaps fit better in our section on that strand of ecological thought. However, there is much deep-ecological writing, from key authors, that would indeed reveal its acceptance of social-ecological tenets, at least from an ecoanarchist perspective (Devall and Sessions 1985).

The review of radical approaches will also require a treatment of the ecofeminist literature, which has become more popular in recent years. Feminism has risen, controversially enough, to the ranks of compulsory learning in all fields of social science, including international relations. And of course it too has many strands, including for example liberal feminism, which is probably closer to mainstream academia today than ever. However, the generic ecofeminist position derives from a concern with both patriarchy and the dominance/destruction of nature, and calls for a radical restructuring of society. As such, we include it in the category of critical literature (see for example Merchant 1980 and Biehl 1991).

AN ECOLOGICAL READING OF IR THEORY: SOME GUIDELINES

As we discuss the key traditions of IR theory from an ecological perspective, we seek to answer two questions: a) do those traditions mention the theme of ecology, and if so, what ecological perspectives do they reflect? b) do those traditions possibly contribute to the process of ecological degradation, through their own ontologies, epistemologies and policy prescriptions?

Answers to the first question will be rather brief, since the prevalent uses of "ecology" in realist and liberal theory are easily summarized and interpreted; basically, mainstream IR theory has been influenced by utilitarian and "Green-Leviathan" ecological thought. Critical IR, on the other hand, makes only passing references to ecology, although one can see how congenial Critical IR is with radical ecology and this is a link that we explore in substantial detail in Chapter 5.

While this first question is interesting in and of itself, it is the second question above which, in many ways, is central to a book of this length. Are the realist and liberal worldviews conducive or anathema to the health of our natural environment? And is Critical IR, in its defense of oppressed constituencies and refined descriptions of global process, also able to "defend nature?" Consider first the "meta-theoretical" questions of epistemology and ontology. Concerning the former, what is the *method of knowing* employed by theories of IR? Mainstream IR theory, as we know, has been secured in the validation of positivism. It has created its own image as a "scientific" discipline, embracing and propounding the thesis that reality can be understood through systematic empirical observation. Many ecologists, in stark contrast, feel that this Western-based positivist conviction has been at the root of modern environmental crises (though, in many cases, they readily acknowledge the potentially constructive role traditional Western science can play in developing temporary solutions and forging the consensus needed to move ahead with longer-term ones).

Next, in terms of ontology, we need to know what kind of world social theories accept as real. This is impossible to describe without some relapse into caricature, of course. But again there are quite obvious distinctions between the two general fields of social theory under investigation here, and significant differences within them. Neomarxists continue to see an international political economy driven by conflicting class interests; realists posit an atomistic world of colliding centers of power; liberals see rational individuals as the most important element in a world teeming with cooperative potential. Ecologists have markedly different perceptions of what, exactly, the world looks like, though they tend to hold in common an emphasis on the fundamental interconnectiveness of all living things. From an ecological perspective, it becomes essential to understand how IR theory views "nature," broadly stated. Ontology and epistemology are obviously linked, since one's view of nature conditions the way nature is observed, i.e. the way knowledge is acquired.

As knowledge is not merely acquired but also used, we then need to consider what paths to a better world are advocated by theories of IR. Robert Cox and others have suggested that social theory has *purpose*, that it aims at producing an effect. The prescriptive purpose of IR theory hinges, of course, on its description of the international political reality, and so we must explain what reality is indeed real to IR theorists, how they use such descriptions for policy purposes, and how such views of the world (as it is and as it can be) may pose a problem (if at all) for the ecological well-being of the planet. Let us remember, of course, that the traditional aim of IR theory is to understand the sources of international insecurity (or disorder) and, therefore, to seek avenues for long-standing security (or, again, order).

Our discussion shows that mainstream IR favors themes that have clear ecological implications. Consider, for instance, some key realist ideas: international conflict is always a possibility; the world is anarchical, the response to which is national hierarchy and homogeneity;

the war effort dictates a utilitarian view of nature. Liberals, for their part, also embrace such utilitarianism as key to progress, assume the universality of their view and encourage global convergence, and are apt to entrust the task of political order to technocrats.

Critical IR theory has already attempted to reply particularly to the universality and hierarchy inherent in the mainstream, key foundations which, in other words, raise anew the debate about the virtues of power centralization vs. decentralization. Decentralization is a mantra of most green thought, but its application to global governance is not clear; the technocratic solutions to environmental problems, lauded as progressive and integrative by some liberals, deplored as manifestations of hegemonic managerialism by some critical theorists, require some degree of centralization. Radical ecology follows Critical IR in this attack on the mainstream, while other strands of ecological thought can presumably be inserted within the mainstream: since the thrust of mainstream IR theory is order and stability, and that of critical theory is change, so can ecological thought display conservative and progressive tendencies.

Admittedly, other authors exploring the nexus between ecological thought and IR theory have insisted on their own list of key concepts. For example, the editors of a popular collection of influential writings on global ecopolitics suggest we "pay particular attention to underlying questions of power, interest, authority, and legitimacy that shape global environmental debates" (Conca *et al.* 1995: 13). These terms are not altogether absent from our discussion, even if we favor others described above. All in all, readers should see this book as a means to uncover the ecological imprint of IR theory and, thereby, to move us closer to what Robert Boardman (1997: 43) has termed "a useful paradigmatic pathway open to creative hybridisation possibilities."

CONCLUSION

The IR literature has mostly treated the environment as yet another issue of collective (in)action, and/or as a trigger of conflict. We are not arguing that such research concerns are illegitimate or irrelevant; much to the contrary. However, we may safely state that IR theorists, by and large, have not explored the wealth of theoretical research suggested by ecological thought. IR theory does not yet recognize the value of ecological thought as political theory. We aim to contribute towards that recognition, one we see as both heuristically formative and, in the longterm, necessary.

We hope to show that ecological thought may be used to both visit and revisit IR theory. Clearly, and as hinted above, ecology has been used as a realist theme – scarcity breeds conflict, and theorists of environmental conflict have adequately mapped out the process by which, say, land erosion or increased flooding may provoke intra/inter-state crisis/conflict.[14] Used restrictively, then, ecological thought may be (and has been) incorporated within realism; indeed, a long line of *geopolitical* writings has always done so. A *critical* use of ecology, however, is more likely to serve as a critique of realist ontology, epistemology and purpose. Again, this is not merely juxtaposing two fields for inconsequential purposes: the normative concerns of ecology and realism do overlap, and as social theories, they both encourage a way to see the world and to construe knowledge.

Similarly, ecological thought may be used as a critique of liberal IR theory. Liberals have also used ecology, treating environmental issues as case studies for theories of cooperation and compliance – and so has appeared the concept of an environmental regime, and resurfaced the statement that natural-resource experts may be/have been empowered to depoliticize contentious international issues (we can trace this back further to the functionalism of David Mitrany) (Mitrany 1966; Long 1993). Explored to the fullest, ecological thought will assess the views on freedom, growth and universalism inherent in liberal IR theory, again stressing their ecological limits. The comparison between ecology and liberalism is particularly interesting in view of liberal commitments to emancipation and material growth.

Finally, we will need to situate ecological thought within critical IR theory. Critical IR makes passing references to "the environment" in its dissection of positivism and traditional conceptions of state and power in IR, yet there is no palpable understanding of the ecological problematique. Utilitarian and reactionary ecology will tend to view critical IR theory as fanciful, if not downright disruptive. This is not to say that critical IR cannot use ecology. In fact, ecological thought itself, at least in its radical form, is already tributary to critical theory (beyond IR), and would seem a logical complement to critical IR – fused within the critical stream, or, perhaps more interestingly, standing out as a specific critical voice, articulating a distinct approach to global power/relations.

We can offer only tentative generalizations here, and the reader has the responsibility of passing judgement on any offered later. The main approaches to international politics that have sought, in very limited fashion, to intellectualize ecological considerations have reduced them to a threat in the realist tradition (scarcity, geostrategic resource war, population imbalance) or an opportunity for material advancement (supported by regime maintenance and institution building) in the liberal tradition. In contrast, the reformist literature (WOMP, peace studies) has been profoundly affected by the gradual recognition of environmental problems, if not by a universally defined ecological crisis; and the critical literature, including that on global social movements, has much in common with radical ecology.

Undoubtedly, ecology as a science has contributed to mainstream IR theory because of the gradual acceptance of biophysical interconnection, which leads and links social scientists who explore economic/social interconnections. This is an intellectual tradition as old as Aristotle. However, we are convinced, and believe this book will amply demonstrate, that ecophilosophy does not mesh as easily with mainstream IR theory, and that their dialogue can only remain a reluctant one. For example, radical ecology demands a vocal and determined prescriptive orientation calling for less economic growth, a paradigm shift in human-nature understandings, and the fostering of peaceful relations between non-hierarchized communities. There is really no equivalent in IR theory, which must deal with the truly staggering global political implications of such a broad and revolutionary agendum.

Chapter 2
Ecological thought
A synopsis

INTRODUCTION

The central aim of this chapter is to introduce students of IR to several of the key themes delineating the more popular lines of ecological thought. It should be clearly stated from the outset that what, exactly, constitutes ecological thinking may be understood either quite narrowly or very broadly. The restrictive view focuses on theoretical developments following the birth of ecology as a natural science, in the nineteenth century; from this perspective, ecological thought is construed as a naturalist philosophy emphasizing the homeostatic character of nature (Hayward 1994: 33), from which ethical and social prescriptions can be derived.

The broader view, on the other hand, stresses the long continuity of philosophical reflections on nature – in the West and elsewhere. The argument here seems two-fold. First, reflective thinkers have grasped at *lessons* from nature, and have proceeded to form the basis of related ethical codes and legal traditions. Second, *worries* about nature are as longstanding as the use of nature itself; in other words, if ecology is simply understood as environment or resource management, then environmental concerns have been clearly central to the history of the world (Ponting 1991), and so equally central to the history of ideas. This is evident within the IR literature in the form of geopolitics.

There is definite value in recognizing this long-standing effort at thinking about/from nature, if only to put in perspective contemporary discoveries about interconnectedness or scarcity. The influence of the natural environment on secular thought was already apparent in Ancient Greece. Philosophy in the pre-Socratic age was not concerned with ethics and politics, but its essence was surely naturalistic:

the pre-Socratics were proto-scientists, searching for some form of order in nature, and in the process speculating on either the essential immutability or the essential dynamism of nature. The early atomists, Leucippus and Democritus, were instrumental in laying the intellectual foundations, through Newton, of a materialist worldview to which many modern ecologists would adhere. While the view of the world as colliding atoms may appear distant from ecology, it is arguably the mechanical principle of interrelated causality that has appealed to ecological scientists – that much is evident from the crucial concept of the ecosystem.[1]

In Aristotle, we find perhaps the earliest references in Western philosophy to some themes which could be called *ecopolitical*, as distinct from ecological. Following the pre-Socratic tradition still vibrant in Eastern Greece, Aristotle was a fond student of the natural world; an early form of science was probably his greatest passion, as he observed and recorded various facts about Greek/Aegean flora and fauna, and as he speculated on various laws of motion. While Aristotle's teleology is ultimately anathema to modern ecological science, and while his treatment of perfection-in-nature differs in many respects from his views on the good life for human beings, he understood that balance (or "appropriate proportion") was an important characteristic of nature which should also apply to human beings.[2] This assumption of balance is at the heart of the Aristotelian view of virtue as a mean between extremes – an argument with decidedly ecological connotations. Applied to the political community, it calls for a limit to population growth which would guarantee the integrity of the Greek polis as distinct from the village unit and the empire. Aristotle's only concern here, in typical Greek fashion, is to ensure that deliberative activity be optimized, and so one should be careful in labelling him as an early environmentalist. Yet there is no doubt that he provides an early lesson in the apparent virtues of equilibrium, or moderation. Excesses are unnatural, and therefore should be discouraged. This argument may well constitute a naturalistic fallacy, yet much of the contemporary ecological literature may be similarly touched by this criticism.

Beyond the secular reflections of classical authors, some distinctive environmental or ecological themes have surfaced in many religious and pagan traditions, as well as in non-Western cosmologies. The concept of stewardship is one such example. Consider also that of intergenerational equity, or the idea that a proper environmental ethic should encompass the rights of future generations; Edith

Brown Weiss has found some of its roots in common law, Islamic law, African customary law, and Asian non-theistic traditions (Weiss 1993). To this day, analysts recognize that monotheistic religions have shown flexibility in incorporating an environmental discourse in their doctrines (Coward 1995), while others also blame them for their fundamental insensitivity to the cause of nature (White 1967).

In sum, a review of ecological thought for a purpose of dialogue or synthesis with IR theory could extend very broadly, and soon become difficult to manage. For this reason, we prefer to circumscribe the discussion to a more established categorization of ecological schools, whose point of departure is traditionally the coining of the term "ecology" by the German scientist, Ernst Haeckl, in 1867. This said, we shall not refrain from examining any pre-dating (yet directly related) theoretical material that will better situate our arguments.

As indicated in the general introduction to this book, we may identify three general schools of ecological thought with direct relevance to political order. The *utilitarian* school includes both the optimistic approach to economic growth characteristic of classical liberals *and* the cautionary warnings of conservationists. The *authoritarian* school stresses the rationalist logic of utilitarianism through a Green Leviathan, or encourages a fusion of ecology with fascism, or argues for a return to a stable, feudalistic structure of political governance. Finally, the *radical* school stresses the links between the control of nature and the control of life, in all its forms. This trichotomy is admittedly Western-centric (Simmons 1993) and echoes some other, more familiar attempts at categorization (namely Eckersley 1992). Before examining each of those in turn, however, it seems essential to pause for a moment and post some guidelines that will help us appreciate, from the outset, the wide variations characterizing the field of ecology as social thought.

"Ecological thought" is not a consensual expression. Ecology is literally the "study of the house," and so ecological thought seems to include sundry reflections concerning the house, i.e. natural habitat. As a science, ecology is a relatively new pursuit, whose objective is to describe the mechanisms binding organisms of appreciable size to their immediate and larger environments (ecology is thus an offshoot of biology, which focuses more on the micro level – although even this differentiation is problematical). Ecological thought, then, hinges on a scientific understanding of natural interrelatedness and balance, expressed in the concept of the ecosystem; it requires an appreciation of nature as more than merely inert matter, of nature – if not alive – as, at least, in motion.

The scope of ecological thought is indeed bewildering, particularly so if the term ecology is used loosely to apply to any sort of literature or activity related in some way to nature. Not all scholars of ecology-as-philosophy will agree with this flexible use of the term, yet it must be recognized that ecology has become a very elastic theoretical device, either masking defenses of the status quo or used critically to spur changes in our rapport with nature, issuing recommendations that will affect our endorsement of accepted political structures and economic practices.

It could be argued that the ostensible purpose of ecological thought is to seek explanations for the sustained degradation of natural habitat and prescriptions for the latter's reversal. As social thought, it is understood that both descriptive and prescriptive elements focus on the human being: human nature, (human) institutions. Of course, a scientific or an impressionistic discourse about natural habitat may be equally the basis of, at times, relatively limited theorizing about the ecological society; and, admittedly, scientists or other empiricists may have expressed some views of nature which were then borrowed by philosophers or other social theorists for their own purposes. Since ecological thought encompasses the various aspects affecting life on Earth, its foundation is necessarily wide.

As a descriptive (or analytic) and a prescriptive exercise, ecological thought articulates worldviews and defends ethical codes, expressed (in part) in political and economic practice. Thus, we ought to expect assumptions concerning the essential place of humans in nature, the value of nature, and the sheer purpose of nature; this is the ontology (and the teleology) of ecological thought, based on some scientifico-experiential reading of nature and either infused of or reacting against enlightenment thinking. The ethico-political prescriptions which follow may or may not be detailed, yet there is little doubt that recommendations for "ecological living" (or sustainability) will logically result from any interpretation of ecological degradation.

Ecological thought is derivative of established currents in philosophy, particularly from ethics and politics. Ecologists have frequently resisted attributions of left or right positions on the political spectrum, yet it seems incorrect to argue that ecologists have really invented a new political front. Both conservatives and progressives abound among their ranks; and so we must cast them within their appropriate philosophical traditions.

We purposely set out to present a wide spectrum of thought that may be called ecological. In the process, we have fused under this one heading two terms which are actually far from synonymous – that of environment and ecology. We hope to clarify this distinction as we present the several key schools. Many established thinkers in the field do consider environmentalism to be anathema to ecologism (Dobson 1995: 1–2), and they may well have a point. This dual usage may be seen as the basis for a series of well-known dichotomies which have captured the rift between the presumed "lovers of nature" and their equally presumed opponents: "value in" vs "value of" nature, ecocentrism vs anthropocentrism, organicism vs mechanicism, holism vs resourcism. These dichotomies are controversial, yet they do succeed in establishing a proper identity for eco-philosophy, i.e. as the purview of a reconsideration of nature as mere useful matter. The result may not always be progressive (and may be downright reactionary), yet it will likely allow ecological thought to offer tangible theoretical alternatives to a mainstream ensconced in contractual exchange, the market, and the problem-solving character of science.

UTILITARIAN ECOLOGY

This section deals with the recognized mainstream of what should be designated as environmental thought: nature is seen as use-value, as capital, to be properly managed – through innovative engineering and judicious use of scientific research in various fields (biology, forestry, zoology, etc.). The approach is typically liberal, i.e. incremental and problem-solving (and likely technocratic). This said, we will need to discuss to some extent what is actually a form of anti-environmentalism, namely the cornucopian perspective, which denies any recognition of environmental problems; cornucopians are not environmentalists (and much less ecologists), yet their now dated (hence extreme) views have inevitably contributed to a fashioning of environmental consciousness in the late contemporary period. Finally, we should mention that a utilitarian perspective also permeates some of the authoritarian approaches to be discussed in the next section. This illustrates the problems associated with social typologies, i.e. the encapsulation of currents which, in the real world, flow from one to the other; this particularly shows how the apparently liberal character of utilitarianism is quite amenable to (if not readily dependent on) concentrations of power.

Philosophical roots

Utilitarianism is an ethical theory with direct political (and ecological) implications. Its original formulation by Jeremy Bentham (1748–1831) legitimized the growth of free-market forces in the Enlightenment and post-Enlightenment era. Utilitarian themes had been already expressed by Adam Smith (1723–1790), whose rejection of mercantilist practices and consequent endorsement of the free market formed the essential theoretical base for the momentous economic changes in nineteenth-century Britain.

Bentham's career was not that of a scholar, yet he was obviously a disciple of Smith and of the Enlightenment project (even if his flexible ethical theory would oppose that of the great Enlightenment thinker, Immanuel Kant). Bentham was above all an active reformer, particularly interested in humanizing the penal system, yet also committed to a liberalization of the political process and of economic laws. His association with the Whig party and with James Mill, who was very active politically on the Liberal side, and his own interests in several financial ventures all shaped his intentions to specify a theory of "the good" based on "utility."

Utilitarianism thus took shape as a consequentialist ethical theory, where the "usefulness" of an act (or a policy) would be judged according to its general, social impact, measured quantitatively in its capacity to "maximize pleasure" and "minimize pain" (Bentham 1952). Bentham's "felicific calculus" has become well known in this respect: pleasures could apparently be measured objectively, according to a series of indicators (e.g. intensity, scope, etc.) and coefficients. Finding the good in a formula is the basis of contemporary social-scientific thought and of technocratic practice. Classical utilitarians and their many followers assume that happiness is above all a function of material well-being: this is secular thought, originating from the Enlightenment defense of human rationality and human rights. As a seemingly democratic principle, it asserts that all (material) sources of happiness are intrinsically equivalent; the point is to measure their capacity to be "enjoyed" by the majority.

The ecological significance of Bentham's ethical theory is readily apparent (and this underscores the importance of evaluating utilitarianism as a social theory, and not as a mere guide for individual action). The pleasure principle requires a treatment of nature as use-value. It does not necessarily entail, in theory, the swift (and unsustainable) ransacking of resources for short-term enjoyment; conservationist practices testify to the possibility of qualifying the "pure" utilitarian view. However, awareness of the "limits to growth" was limited only to a tiny minority, in Bentham's time. And so, in practice, the pleasure principle has become a licence for extracting immediate use-value from forests, oceans, and animals alike; restraints may be placed by governments responding to minority interests (showing scientific credibility), but the logic of competition (which is key to the modernist ideology in which utilitarianism is imbedded) is arguably overpowering.

This said, utilitarianism is not altogether inimical to some form of an environmental ethic which would qualify the pleasure principle. On the one hand, we must recall that animal-rights activism originated in Bentham himself and in various progressive policies in nineteenth-century Britain. Bentham was well-known for his love of animals, and the pleasure principle can logically be extended to animals as sentient beings (Singer 1975); as Bentham pointed out, animals do suffer. However, Benthamite utilitarianism is quickly faced with a paradox when the criterion of sentience is applied to its logical conclusion, i.e. when one points out that suffering is an inevitable byproduct of resource extraction on a large scale: animals of various size and appearance directly suffer when in contact with pollutants, harvesting machines, the factory floor and cage, or the scientist's lab; meanwhile, countless human beings have suffered from the pressures associated with global capitalist growth. In other words, Bentham's utilitarianism does have an environmental component, i.e. it may be used as a way to advocate restraint against the assault on nature. But ultimately, while Bentham refuted Descartes' treatment of non-rational life as essentially "dead matter," he did not shed the philosophical dualism that inexorably leads to an elusive mastering of nature.

Beyond the issue of animal rights, an argument linking utilitarianism to environmentalism can also be defended if the ethical theory is understood not from Bentham's perspective, but from that of John Stuart Mill (1806–1873). Mill's stature as a philosopher is scarcely matched by Bentham, yet he is obviously indebted to the latter for constructing a political and an economic philosophy based on the ethical theory of utilitarianism.[3] The key to Mill is his revision of the pleasure principle, shifting its ground to a qualitative assessment of happiness. Mill's utilitarianism must be understood from the perspective of a subtle thinker concerned about the cultural vacuity of a democratic society in full industrial revolution. Mill understood the homogenizing impact of a politico-economic system based on simple majority views and on industrial growth. The principle of "the greatest good for the greatest number" encouraged both the "tyranny of the majority" (hence the marginalization of minority views) so perceptively discussed by Alexis de Tocqueville (1805–1859)[4] and the proliferation of consumption goods supposedly enhancing societal happiness. Happiness had become the mass enjoyment of low-level pleasures, conditioned by the interests of the growing entrepreneurial class and the expanding forces of production which they encouraged.

As a liberal who believed in the importance of private property and personal rights, and as a modern who resisted the encroachment of religious authority on the reasoning being, Mill could not shed the basic principle of utilitarianism – which seeks the gratification of the individual here on Earth. Yet Mill was also a romantic and, above all, in typically Aristotelian fashion, a thinker who resisted any facile application of rules. From an ecological perspective, his revision of Benthamite utilitarianism is important, as it formed the basis of recommendations for a more sustainable relationship between humans and their natural environment. Mill emphasized the quality of "higher-level pleasures," to be attained through critical reflection and through the contemplation of nature.[5] Here, and notwithstanding charges of elitism which he rejected,[6] Mill argued that mass consumerism impoverishes the mind and the Earth, and enslaves while pretending to free. The economic, political and ecological implications are all clear: freedom hinges on a respect for minority views, some redistribution of wealth through government intervention, worker management and ownership, and the establishment of a steady-state economy after a period of moderate growth (Heilbroner 1961: 109–10). In his defense of freedom, Mill had qualified the blunt liberal discourse by incorporating lessons from Tocqueville, romantic poets, and also Thomas Malthus (1766–1834), who had explored well before him the physical limits to population growth.

As students of ecophilosophy and ecopolitics, we are then confronted with the realization that utilitarianism *à la* J. S. Mill is quite different from the rather crass version of Bentham. We could perhaps address this paradox by arguing that Mill was not a real utilitarian – that, if Mill had experienced ozone depletion or radioactive fallouts, he would have concluded that utilizing nature through the forces of modern science and technology is likely to be self-defeating. On the other hand, Mill had ample trust in the power of reason, in the human ability to transcend petty interests and properly manage any enterprise for the common good. And so Mill can remain a utilitarian in the restricted sense, and allow the ethical theory some ecological credibility. In other words, a Millian form of utilitarianism may stand within a materialist tradition, where rules of ethics are imposed neither by religious/traditional authority nor by a categorical imperative, while resisting the historical implication of a materialist worldview, i.e. turning matter into disposable resources.

Utilitarian anti-environmentalism

The contemporary utilitarian perspective is more akin to a technocratic political position that explicitly rejects calls for linking environmental and social problems. In spite of massive evidence pointing to environmental decay, classical utilitarian assumptions about and policies for nature are still held by a sizeable minority of thinkers with substantial political clout. Their minority status is in itself remarkable, considering that the unqualified Benthamite view of nature held sway until the early 1970s.[7] Throughout the nineteenth century, and as reflected in liberal IR theory, the mainstream gathered around assumptions of unlimited material progress; conservationism was an eminently progressive movement at the time. This basic mind-set, although increasingly challenged by the odd conservationist treaty or law, endured through most of the twentieth century. It has become known as "frontier economics," a belief in the inexhaustability of nature reminiscent of the early days of the North American pioneers.

As we stand near the third millennium, there remain some voiceful opponents of even the mildest form of environmentalism. Wallace Kaufman, for instance, recently published a book denying the existence of an environmental crisis, suggesting that the environmentalist movement is merely reflective of some psychological trauma, some deeply felt angst in a world of modernity (Kaufman 1994). From a different perspective, we can point to a rather extensive scientific literature dismissing the environmentalists' claim, i.e. that of crisis or impending crisis (Baarschers 1996, Bailey 1995). In an obvious response to the Worldwatch Institute's "State of the World" publication series, Robert Bailey rejects the environmentalists' "precautionary principle," arguing that

> It is impossible to know all the consequences of the most trivial action . . . It is better to move forward using intelligent trial and error to uncover new knowledge . . . and wealth. Greater knowledge and wealth give human communities resilience, enabling them to respond flexibly and effectively to the unexpected.
>
> (Bailey 1995: 5)

Authors such as Bailey, and especially Julian Simon, have defended the human capacity to solve any problem related to matter through reason, through human ingenuity (Ray and Guzzo 1993; Easterbrook 1995; Simon 1996). And so they have built on the initial criticisms of the 1972 Club of Rome Report, dismissed as hopelessly alarmist (Adler 1973; Beckmann 1973; Vajk 1978; and, more subtly, Cole *et al.* 1973). It is not that these thinkers endorse the naive cornucopianism of classical liberalism; rather, they invoke technical solutions to problems of production (including "negative externalities") and, ultimately, technocratic rule for society (Simon and Kahn 1984). By calling upon the power of engineering, they contend that scarcity is a problem only in principle, and not in practice.

In sum, while it is still fair to argue that most scientists, engineers, economists, and lay people have developed some environmental ethos recognizing the need to recycle and to "use wisely", there remains an articulate minority seeking to bypass such restraints altogether. They can be labelled as "anti-ecologists" or "anti-environmentalists" in the absence of any critical argument that would side with environmentalist concerns. Their critique, in fact, pertains to the inefficiency of a technological system which, presumably, can be much improved. Arguably, their views ultimately lead to the concentration of power in the hands of corporations, which can hire the specialized labor and pay for technical education necessary to sustain such ecomanagement.

Conservationism

Conservationism is the most recognizable form of ecological thought today, although it is more readily associated with environmentalism than ecologism, given the radical connotations of the latter. Conservationism is reformist thought and practice, legitimate in the eyes of governments and the public alike (at least in the West). It remains utilitarian thought, moving away from Bentham yet not quite reaching Mill's romanticism. Essentially, it calls for restraints on natural-resource consumption so as to respect the "carrying capacity" of ecosystems. The motto is prudence, based on a scientific understanding of ecosystemic balance; yet this in no way forbids science and technology to enhance natural-resource yields, if they can demonstrate such ability. Ultimately, capital accumulation is not seriously questioned. The fracturing of nature for the purpose of assessing yields detracts from its holistic character, and thus sustains the classical liberal confidence in productive growth. Conservationist language mixes well with apparently innovative, market-based solutions to problems of collective action (Brown *et al*. 1996).

The origin of conservationist thought may be found in the latter half of nineteenth-century America, and here conservationism must be distinguished from a more radical, preservationist attempt at environmental protection (Norton 1991: 7–8). Preservationism is not a managerial philosophy or policy; it may not be a completely holistic form of ecological thought, yet it does attribute inherent value to species and ecosystems, and so will be discussed later on.

Preservationism and conservationism formed twin reactions to the impact of economic development on nature in the US. As Norton explains, conservationism emanated from government, concerned specifically with the growing depletion of forests. The burgeoning science of ecology would soon stimulate interest in forestry: ultimately, the first Forestry Department would be created by the American federal government and headed by a renowned scientist, Gifford Pinchot, whose general mandate was to ensure the long-term sustainability of forests for purposes of economic growth (also Attfield 1991). In other words, conservationism emerged as a governmental reaction to rapid growth, and would pave the way for extensive environmental legislation and regulation in the twentieth century. However, no fundamental challenge was posed to the power of instrumental rationality or the established ideology of growth: this was management for the long-term good of the public, the state, and capital (which did not know better and which would adapt very quickly, once persuaded by scientific language).

The innovative character of conservationism was to use new research in emerging fields of natural science so as to make socially credible arguments in favor of environmental management. However, questioning about carrying capacity or maximum yield did pre-date the movement. Mill's romanticism cannot be fully reconciled with the problem-solving mind-set of conservationism, but his work on economics does yield some support to conservationist policies. Although not a scientist, Mill understood that economic logic was restricted to production,[8] and that the productive capacity of land was finite.

As mentioned, Mill was influenced by Malthus, whose pessimistic visions for a sustainable future may be conjured at this juncture. It is not clear whether the writings of Malthus or Mill did influence nineteenth-century conservationism an ocean away. Furthermore, Malthusian themes are often invoked in analyzing authoritarian approaches to sustainability, and so Malthus should perhaps be considered in our next section. Nevertheless, he is by no means out of place here.

Malthus, too, was an economist, and his focus was on the ratio of food to population in England, which appeared to be diminishing in the late eighteenth and early nineteenth centuries, and which presaged famines and ensuing social disorder. He could not have predicted the advances in scientific knowledge and technological prowess which would seemingly push back indefinitely his apocalyptic scenario, and so he remained a marginal footnote to economic history for some time. The point, however, is that a conservationist logic was at least implicit in his work. The only economic factor that he assumed controllable was population, and so his recommendations specifically aimed at population control, rather than the management of natural resources. Yet the conservationist logic, at its base, pertains to the management of human consumption, i.e. to the ratio between available natural resources and mouths to feed, or houses to furnish, or children to dress. In this, Malthus is in familiar territory, and this explains particularly why his name and his doctrines have resurfaced as part of twentieth-century mainstream environmentalism – in both democratic and authoritarian discourses.

The decades since the Second World War, and particularly the last quarter of the century, have witnessed many expressions of conservationism as environmental thought. It is fair to say that the conservationist discourse is much more refined than in the time of Pinchot. The utilitarian roots are still solidly anchored, yet the defense of sustainability has broadened into time and space. In other words, mainstream environmental language still insists on economic development, but does qualify it with calls for redistribution, efficiency, and moderate restraint. Overall, one can see variations in the degree of reformism advocated by the various groups using environmental language in the political arena. Our objective, then, in the next few paragraphs, is to identify how contemporary conservationism has been expressed, and to point out the shifts in theory from a purely managerial ("resourcist") approach to outright preservationist (yet not "radical") advocacies in selected areas.

Basically, the analyst may investigate three locations where conservationism is expressed today: in government literature and policy, in NGO advocacy, and in key policy-oriented publications by independent and semi-independent authors or commissions. Again, the challenges to the economic status quo will vary from one location to the other (and within each), yet there is an undeniable convergence between the three which still manages to steer an "acceptable," middle road.

We may begin with the key publications, where, after all, a theoretical base is most likely to be articulated. Important contributions were made particularly by Paul Ehrlich (1970) and Barry Commoner (1972, 1976). Ehrlich's demographic studies played an essential role in rekindling the Malthusian fears of a population bomb, and his writings were to be used both by progressives and conservatives – i.e. either to encourage a moderation of consumption in the North or to urge the imposition of drastic birth-control programs in the South (or even forceful neglect of the poor and famished masses). Commoner is less easily categorized here, for his many conclusions emanating from his research on thermodynamics and ecology are strongly critical of capitalism. Nevertheless, Commoner established himself as an author for the general public, and a close reading of his works points to a clear anthropocentrism which may not suit the radical ecological wing. Like Ehrlich, Commoner has had a direct impact on government, but above all as a scientist and not as a social theorist or critic – and that is the key to his inclusion here.

In addition to the above two authors, several path-breaking scientific studies stirred environmental consciousness in the West, particularly Rachel Carson's *Silent Spring* (1962). A book such as Carson's, however, led to radical conclusions for which governments and the general public were ill-prepared. Accordingly, its concrete impact was to nudge governments and the (American) public away from classic, no-holds barred approaches to economic growth, and to lay a small stone in the path towards the reform environmentalism of the 1970s. In other words, it would be incorrect to consider Carson's book as an example of conservationist thought, since her scientific discoveries called for much more than proper management or proper use of natural resources. But its impact in stimulating conservationism is undeniable.

While many other scientific or scientifically-laden works could be cited, it is the material contained within two commissioned reports which has drawn most attention in any review of contemporary eco-political thought. These two reports are, of course, the Meadows report to the Club of Rome (1972) and the Brundtland Report to the United Nations (1987).

The Meadows report arguably launched the current wave of reform environmentalism, seeking, through legislation, responsible use of natural resources and opening the door for more clearly preservationist practices (see below). Proper resource management is key to the report's recommendations, in view of the inescapable logic of exponential growth. The report is clearly in the line of Mill and Malthus, both of whom are cited on several occasions. Computer models seemingly demonstrate the need to even out global birth and death rates, in this case by extending Northern patterns of demographic stability to the South. Emulating Mill, the authors do advocate a steady-state economy which, however, would not necessarily forestall progress in the quality of life. Technology is by no means a forbidden word, and so it would be applied to various programs of recycling and environmental management.

In sum, there is a recognition of a long-term price to progress, i.e. a Malthusian warning adapted to a new technological reality. In a reform-liberal, conservationist fashion, the report is pessimistic towards existing patterns of growth, yet not necessarily so towards the likelihood of adjustments to the system of production. As a submission to an elite group including (particularly) business executives and scientists, the report shies from any recommendation that may directly question the legitimacy of global capitalism. Invoking Herman Daly, the authors do hint at the need for redistributive policies that would favor equality. Yet the language ultimately settles for a rather vague urging (quite technocratic in tone), "from [their] vantage point as systems analysts", to "thoroughly analyze" the "underlying structures of our socio-economic system" (Meadows et al. 1972: 186).

The Meadows Report could thus be appreciated for its careful data projection technique, echoing and strengthening the credibility of an old message about population-to-resource ratios. While radical solutions could have been read into the report, most radical critics would see it as an alignment with the mainstream. At the very least, the report raised the necessary global consciousness for a reappraisal of growth patterns, nonetheless steering clear of any political controversy.

The Brundtland Report would take a clearer political stance concerning the future of the planet. Brundtland represents, in fact, the culmination of fifteen years of global assessment reports, which included the Brandt Commission ("North–South") Report of 1980. The Brandt Report is not known as a main signpost in the recent history of utilitarian environmental thought; there are more than passing references to the environment, but, in the obvious spirit of the New International Economic Order, the report stresses, above all, the various ways to spread the fruits of economic development to the Third World. Scarcity (as fact and possibility) is recognized for the South and for the world, and so the usual techniques of ecological redress are mentioned: birth control, alternative energy sources, waste management – in a word, "rigorous conservation" (ICIDI 1980: 277). This goes along with full-scale investment in the Third World, including in oil and hydroelectricity.

Brandt (an "economic" report) struck the chord of equity which only vaguely resonated in Meadows (an "environmental" report).[9] "Economics" and "environment" converged in Brundtland, which popularized the concept of sustainable development, a mainstay of utilitarian ecological thought. As expected, the idea of a "sustainable development" was promptly denounced as contradictory, and therefore misleading, by many critics (Lélé 1991). There is no doubt that, in using such an elastic term, the report's authors sought both to pacify the environmentalist center-left and to retain the support of the business establishment for some effort at resource management. The report is indeed quite insistent in its coverage of serious environmental problems affecting the planet, while pointing out that poverty is eminently linked to environmental degradation and might pose an obstacle to binding international environmental agreements. And so the idea is to develop more cautiously, to resist any obscene pillaging of nature for short-term gain, to open up trade and financial opportunities for Southern manufacturers, and to instill a measure of efficiency in the world economic system that would (obviously) reduce waste. The report recognizes globality, and arguably ends up treating the planet as one integrated polity, for which efficient economic development policies are naturally recommended.

The liberal, utilitarian, conservationist trend of ecological thought is clearly apparent in a defense of sustainable development. The Brundtland Report is careful to recognize the rich cultural (and biological) diversity of the planet, yet the tone is arguably universalist, and unarguably committed to the rational use of nature. Sustainable development simply extends the ecological threshold of no-return, and keeps alive the liberal argument that technological innovation and wise use (at reduced global population levels, of course) will respect the carrying capacity of the planet.

The environmentalism of appropriate management is also the purview of much environmental law and, to some extent, of the positions defended by well-known environmental groups. Both of these deserve a rapid look, since they generate much of the literature devoted to a reform of anti-environmental practices. Some of the most globally influential environmental NGOs, such as Greenpeace or Friends of the Earth, advocate changes to the system of production and consumption that are actually much more radical than what is commanded by conservationism, and so should not receive treatment here; such groups (particularly the more centralized Greenpeace) have taken stances that directly attack the assumption of nature-as-commodity, developing a discourse tying ecology to peace and social equity. Other groups, such as the World Wide Fund for Nature, have fought more specifically for the preservation of biospecies, and so while their language is more technical (scientifically) and less socially explicit, their preservationism goes beyond the conservationist idea of wise use and the conservationist alliance with the private sector.

This said, several other NGOs do come closer to the conservationist stream, playing an essential role in introducing conservationist laws and policies and shaping the accepted mainstream of an environmental ethic: nature must be used, market forces and state (regulatory) power must both be harnessed in fashioning appropriate "environmental incentives," and some areas of nature should be fenced in or otherwise protected so as to allow their regeneration and future exploitation. The Worldwatch Institute and the World Resources Institute (WRI) in the US are two noteworthy examples. Both have been praised for awakening governments and the general public to some disturbing facts about the state of the world. Worldwatch's yearly reports on the state of planet Earth are now translated into 27 languages, and are required reading for many Fortune 500 CEOs and members of the American Congress (Brown *et al.* 1996: xvii). The 1996 edition was branded (on its cover) as a "report on progress toward a sustainable society" – language familiar to readers of the Brundtland Report. Among the several chapters were two dedicated to "sustainable industries" and "market environmentalism," sending a clear message as to Worldwatch's "practical," conservationist approach to a healthy planet.

The WRI, similarly, has pursued elaborate scientific analyses in the field of ecology, providing credible grounds for a less callous treatment of nature by industry. A former vice-president of WRI, Jessica Tuchman Mathews, acquired substantial fame within policy and academic circles in the 1980s by arguing for a "redefining" of the concept of security that would include environmental threats to health and property (Mathews 1989). As a research and advocacy group dedicated to a more "rational" use of natural resources, the WRI legitimately receives the conservationist label. It may not be unaware of the irretrievably social aspect of environmental degradation, yet its reformist approach emphasizes conservation and wise use, as best demonstrated in its pivotal contribution to action plans for tropical forestry, in association with the World Bank. The WRI has been indeed criticized for this, ostensibly protecting biodiversity so as to "increase [its] utility" (Schücking and Anderson 1991: 32; also Gray 1991: 64).

NGO conservationism, furthermore, is perhaps best revealed in its launching of the debt-for-nature swap in 1987. This is a mechanism of debt alleviation by which conservation groups purchase portions of a Southern country's national debt at discounted prices, and offer to write off that portion in exchange for a government commitment to maintaining the integrity of a natural reserve (Page 1989). Although there is undeniable ecological value in the creation of national parks, critics see limits to such Northern intervention. Debt-for-nature swaps do not tackle what presumably drives the ecological crisis in the South: skewed patterns of land tenure, intricate dependence on Northern markets, capital, and development models and, as a consequence, profound marginalization of newly urbanized masses and indigenous groups. Thus, they are often criticized for luring Third World governments into lower-priority environmental policy and, as well, for displacing forest dwellers from reserves. In sum, the swaps seem to represent, at best, a timid approach at preserving biodiversity, or, at "worst," a bona fide conservationist measure securing the long-term utility of natural tracts, particularly tropical forest patches, to be harvested "sustainably" (Gray 1991: 62–3; see also Mahony 1992).

It would seem more than difficult to provide a genuinely comprehensive assessment of the ecological orientations of NGO environmentalism, in both North and South. And so the point is more to demonstrate that most of the recognized groups which have commanded big budgets and have had access to policy makers can be classified as conservationist, and have contributed to a legitimization of the utilitarian discourse inherent in conservationism. There are cases where the alignment with the forces of capitalist growth are more obvious than others, as may be attested by the endorsement of continental free trade by the former President of the National Wildlife Federation (Hair 1991). On the other hand, several other renowned wildlife groups, such as the Audubon Society or the Sierra Club, would appear much more cautious on this issue, showing a preservationist commitment worthy of a John Muir or an Aldo Leopold.

Conservationism is, finally, most obviously expressed in various environmental acts of legislatures (particularly) around the Western world. In the wake of (and following) the United Nations Conference on the Human Environment in 1972, and in response to the Meadows report, the American government took noticeable initiative in conservation legislation by introducing the Endangered Species Conservation Act of 1969 and the more muscular Endangered Species Act of 1973; the latter followed the Clean Air Act of 1971. Around the world, environment departments and ministries were appearing for the first time within government administrative structures in the early 1970s. Since then, celebrated attempts at "greening" economic practises have been encountered in several countries, particularly in the early 1990s when a wave of environmental awareness leading to the 1992 Rio Summit spurred governments into action; the Netherlands' National Environmental Policy Plan (1989) and Canada's Green Plan (1992) are two particular examples. At the intergovernmental level, the United Nations Environment Programme (UNEP) emerged as a celebrated result of

...er international environmental agreements would ...s the Convention on International Trade in ...cies, the Basel Convention on Toxic Waste Trade, ...treal Protocol on Ozone Depletion, and the post-Rio framework agreements on biodiversity and climate change.

Obviously, these measures should not be seen as radical reconsiderations of the production system. They have targeted some excesses by notably protecting some well-recognized and loved species of mammals, controlling smog build-ups which pose direct health concerns to (largely northern) urban dwellers, and attempting to phase out a small number of chemicals diminishing Earth's natural protection against deadly UV rays. The ultimate objective, however, is neither to slow down production of industrial goods, nor to redistribute economic or political power, nor even to institute some general commitment to animal rights.

Assessment

Utilitarianism offers the foundation of an "instrumental ecology." As ecological thought, it has crystallized the view of nature as resources for purposes of capital accumulation, and so has created an "environment" around human life – an environment to be approached prudently, but decidedly by the related forces of industry and finance. This is conservationism, a problem-solving approach to environmental crises; it emphasizes improved management, improved environmental technology, and improved compliance with existing international treaties. Conservationists seek to save certain areas of wilderness, to be sure; but these are cordoned off from the larger and inexorable path of human progress in modern society. Conservationists are aware that vast engineering projects have precipitated extreme losses of biotic richness, leading to the fall of empires (Catton 1980; Ponting 1991); these must be avoided through better management, and not through a fundamental challenge to liberal (and neoliberal) value sets.

Self-described pragmatists will value such environmentalism. Provided one is willing to work with authority, incremental progress is possible, especially in Western democracies where governments can be influenced by popular pressure. Conservationists are in no way seeking a fundamental rethinking of the relationship between humans and their natural environment; however, steps must be taken to counteract the potential excesses of a rapport based on commodification. Radical approaches will be brushed off as hopelessly "romantic."

Naturally, such incrementalism receives less than affectionate attention from radical green thinkers. Arne Naess (1972) and Edward Goldsmith (1988) see in this a "shallow ecology," which, in effect, has facilitated the co-option of the environmental movement by business and big government. We have thus seen the gradual professionalization of environmentalism, turning environmental groups into profitable career paths for lawyers, accountants and engineers. Conservationism can be promptly used or trivialized by corporate interests, artfully hiding environmental skeletons behind green marketing strategies, and alarming radical critics with their ability to penetrate further the policy-making circles of governments with sustainable development platforms. In the early 1970s, some Marxist authors actively criticized "reform environmentalists" for their connections to corporations (Ridgeway 1970; Weisberg 1971).

Utilitarian ecology may thus be charged with supporting some new form of global managerialism (Sachs [ed.] 1993). A central criticism of such managerial efforts is that environmental problems are not, in general, managed at all; rather, they are shifted elsewhere. Paul Wapner makes this point with three examples: Japan's insatiable timber demand, which has shifted Japanese forest consumption to other states such as Indonesia and Canada; the Florida trade in coral, which relies mainly on (illegally obtained) coral from reefs near the Philippines, while Florida's reefs are well protected; and the more complex yet perhaps also more disturbing export of dangerous pesticides from the North to the South, despite the banning of the chemicals in the area of production and in spite of the Basel Convention (Wapner 1997; also Dryzek 1987: 10–11). Others argue that managerialism is simply the continuation of a system of North–South exploitation; it does not vary from the dominant development discourse, insisting environmental problems can be solved by "capital, bureaucracy and science – the venerable trinity of Western modernization" (Sachs 1992: 35). This lends itself to an "ecocratic" approach that avoids democratic input (see also the articles in Brecher *et al.* 1993).

As will be explored below, much of the IR literature shares the philosophical basis and the policy outlook of utilitarian ecology. Writings on international environmental affairs are replete with investigations into the process of problem management, centering around the creation and maintenance of regulatory multilateral regimes at the state level of analysis. And, just as importantly, the liberal IR literature adopts perspectives on peace, security and freedom which are evidently based on utilitarian conceptions of nature: the control of nature through cooperative scientific management, facilitating the extraction of "value," is essential to a liberal world order.

AUTHORITARIAN ECOLOGY

The utilitarian stream of ecology, as discussed above, treats ecological degradation as an externality of production, in the liberal quest for freedom through capitalist growth. Liberals will not necessarily reject the coercive power of the state as a means to dissuade ecologically harmful practices. However, preferred means include education campaigns; market-based incentives (e.g. tradable pollution permits); and, of course, applications of scientific principles and new technologies of conservation, encouraged by the state and/or willingly purchased by corporations (so as to cut long-term costs and/or respond to "green consumers"). All in all, for liberal utilitarians, ecological degradation is not to prompt any fundamental reconsideration of authority patterns or of the key values to be pursued in modern society. "Healthy nature" is not an end in itself; and so freedom of exchange, individual (and corporate) performance and innovation, and popular sovereignty are all absolutes to which "nature" is subordinate.

In the last section of this chapter, we will examine how some strands of ecological thought may argue that much of the emancipatory mission of the liberal project may be retained without dismissing nature as accessory, or use-value. For the moment, however, we must explore how ecological thought may also become authoritarian thought and even reactionary thought, within and beyond a utilitarian framework.

Ecology and the Green Leviathan

Invoking centralized power as a means to secure the common good is a time-honored prescription in political theory. Plato first articulated the argument (in the West), handing the reins of the model state to an intellectual elite whose capacity to understand absolute truths would presumably guide a stratified society through the vagaries of life. The just society, serving the common good by maintaining long-term social stability, would be free of corruption and thus hinge on a dispassionate and disinterested approach to rule by the elite: centralized power entailed wisdom and austerity for the ruling class, a communitarian life of physical detachment and intellectual prowess, an enduring control of passions and base needs by reason.

Since Plato, the virtue of centralized power has been most famously exposed by Thomas Hobbes. Of course, there is a world of difference between the Greek pursuit of justice and the seventeenth-century quest for security and material progress, although there are obvious Platonic echoes in Hobbes' own pursuit of objective truths. Hobbes' view of nature (including human nature) was mechanical; fascinated by the path-breaking work of scientists and mathematicians, Hobbes learned from Galileo's cosmology and from Descartes' dualistic conception of mind ruling over matter. Hobbes assumed a basic freedom characterizing human beings in the so-called state of nature, i.e. in the absence of effective government; such freedom was dangerous to the average human being moved by the fear of death, and so a secure life required ordered relationships, arrived at rationally. As a contract theorist, Hobbes turned Greek theory around by treating the political community as an artificial entity, yet a necessary achievement in the path to long-term social stability and individual security.

Hobbes' projected outcome of the social contract is well known. Individuals would rationally agree (through their own will, their own deliberation) to a devolution of sovereignty, relinquishing the right to govern themselves to a Leviathan, i.e. a powerful governing body combining legislative and executive powers. The relationship between state and society would be a fruitful one. Citizens would be quite free to pursue individual interests as long as national security was preserved; entrepreneurial freedom was particularly recognized, as capitalist growth could never have been construed as a security threat, at the time. As the Leviathan is a creation of the people and acts upon its "one will," and as it is shielded from petty considerations that would characterize the Lockean alternative (i.e. a system of representative democracy based on the separation of powers), it is less likely to degenerate towards governmental corruption or any other failure of duty.

Hobbes' Leviathan thus represents the modern defense of centralized power, aiming at the swift implementation of policies designed to protect the common good in an artificial congregation of self-interested, naturally clashing individuals. The state, or commonwealth, is the artificial body built so as to regulate the colliding impulses of natural bodies, i.e. human beings.[10] Hobbes' theory does not emphasize a Platonic intellectual wisdom, and it construes the state purely in defensive terms (and not in teleological terms of moral growth). However, it offers a model which could lead to the empowerment of a "knowledgeable" elite (backed by military force) in the pursuit of national security. If ecological degradation, as a byproduct of "natural" economic competition, can be conceived as a security threat, then one may well invoke a Hobbesian solution to the threat – i.e. a strong governmental authority, based on scientific knowledge, decreeing the "ecological society." This would be indeed a "Green Leviathan."

Centralizing solutions may well accommodate utilitarian worldviews. Environmental management shifts from market incentives to extensive environmental policy and technocratic rule, but the "utility" of nature is not necessarily denied; and so authoritarian approaches to the ecological society may well be modernist, even if skeptical or downright critical of the private sector's capacity to behave "responsibly."

This said, there is a decidedly reactionary tone to Green-Leviathan literature. The best-known advocate of "ecological imposition" is surely biologist Garrett Hardin, who earned his credentials by writing tightly argued books and articles showing the "necessity" of authoritarian measures. For Hardin (and for liberals), ecological survival entails avoiding (or surviving) the so-called tragedy of the commons, whereby "rationality" dictates an inexorable depletion of common resources (e.g. a pasture, or even an oil field) through gradual repudiation of collective-action principles: without a watchdog, and beset by the logic of competition, it makes short-term sense to increase one's exploitation of a commons.

The commons is thus a state of nature, requiring government. The Lockean solution would have the commons formally divided and managed through the allocation of property rights, and Hardin would agree. However, not all commons problems may be thus handled. On the one hand, common sinks (e.g. bodies of water, the atmosphere – thus distinct from common resource pools) cannot be fenced in; yet, admittedly, market mechanisms may still be invoked here. On the other hand, the market is not likely to maintain a proper resource-to-population ratio. In the case of a finite number of herdsmen exploiting a pasture or of companies emitting sulfur dioxyde, liberal solutions may be applied. But when the number of consumers (or depleters) increases exponentially, in the presumed absence of any genuine human altruism, the state must get involved for purposes of redistribution or forcible restriction. Hardin, of course, is well known for defending the latter, for advocating a "lifeboat ethics" that surely too crudely reduces planet Earth to a sinking ship unable to carry all its passengers. As crisis situations command swift decision making and cold-hearted choices, the neo-Malthusian ethic orders that the wealthy and privileged survive and the poor and hungry be sacrificed (Hardin 1974).

William Ophuls is another author whose centralizing arguments have been particularly cited (e.g. Orr and Hill 1979: 310). Ophuls has persuasively defended the centralizing thesis, unfazed by its elitist character: the management of scarcity requires a combination of brainpower and deterrence, which only an alliance between science and state can provide (Ophuls 1977: 157). Ophuls thus recognizes the power of Hobbes' logic, even if his own recommendations for an ecological society are not straightforwardly modernist: they alternatively evoke Rousseau's communitarianism, Mill's steady-state economy and cultural diversity, and a return to an aristocratic system of government that would decidedly stamp out egalitarian tendencies (Ophuls 1977: 226–32).

All in all, Rousseau is probably Ophuls' main inspiration, considering his commitment to direct democracy in a small community setting (Ophuls 1992). However, as a practical reformer, he would dismiss Rousseau's belief that government heavy-handedness is useless as a means to rehabilitating a decaying social body (as opposed to preventing decay). State power is a solution, and the only logical solution. Ecological behavior can presumably be imposed, by a caring, strong and honest paternal figure. Ecological soundness would appear reducible to resource abundance; and while Ophuls might object to this reading of him, his views may not necessarily guarantee the basic rights defended by liberal thinkers. Far from upholding a social contract, Ophuls' authoritarianism may rekindle (false?) hopes for a stable feudalism, where tight social hierarchies could ensure the long-term maintenance of ecologically responsible behavior. In such light, ecological thought may well be construed as reactionary thought.

In sum, and in contrast with conservationist approaches, this strand of ecological thought clearly articulates a political solution to the ecological crisis; in fact, to invoke the swift powers of a Green

Leviathan is surely to recognize that there is indeed a crisis, requiring rapid reaction. Green-Leviathan discussions arguably do not provide any new insights on the essence/meaning of nature, human and non-human. There is no new metaphysics or ontology, and, for that matter, no revolutionary rethinking of politics and economics. When the famed historian, Arnold Toynbee, writes that scarcity will stimulate "within each of the beleaguered 'developed' countries [. . .] a bitter struggle for control of their resources", leading inevitably to the imposition of authoritarian regimes,[11] he echoes a familiar message (see also Heilbroner 1980). We would not characterize Toynbee as an ecologist, yet Green-Leviathan approaches underscore one basic point, i.e. that "environmental concerns" can easily be equated with scarcity, and that solving problems of scarcity has traditionally borrowed the path of centralization. The ecological language is eminently utilitarian: nature as resources, as use-value. But the political language is decidedly not liberal.

Ecofascism

As judged from the above, environmental security may be construed as a social goal for which state authority is necessary. A Green Leviathan seeks social stability through authoritative resource management, and thus emphasizes efficiency and order. Hobbesian ecology is the politics of scarcity, to use Ophuls' terminology.

In fascist ecology, we find a very different understanding of the links between authority and nature. The Nazi regime has been analyzed for its apparent commitment to the harmonization of society with nature (Bramwell 1989; Biehl and Staudenmaier 1995). Ecofascists were not mere environmental problem-solvers. Nazism defended an ideal of nationhood based on agrarianism and the "natural" superiority of the Aryan species, themes that began spreading within Prussia over a century before. Vegetarianism and organic farming were favored by some members of (and structures within) the Nazi establishment; peasant life was revered, while forest preservation was heavily emphasized.

Nature was also admired for its sheer power: this is the other aspect of the blood-and-soil ideology, stressing the survival of the strong and the ominous fate of the weak. Much of the Nazi propaganda thus echoed nineteenth-century calls in Germany for the cleansing of the social body through war, calls merging with a form of social Darwinism whose influence clearly reached beyond the British intelligentsia. Roy Morrison refers to this as a "redemptive authoritarianism" for the sins of industrial modernity. The state is viewed as the most appropriate conduit for the power needed to turn back the clock on the changes resulting from the industrial revolution; power is asserted against change through the medium of mythic nationhood fused with the modern state. In the grand style of fascist epic, this involves nothing less than the "creation of new men and women, not simply new political orders" (Morrison 1995: 114).

In other words, ecological thought is also the purview of a German idealist tradition which very much influenced nationalist and romantic historians and philosophers. The association between, on the one hand, the austerity and grandeur of nature, and, on the other, the violent pursuit of heroism, is of course characteristic of Nietzsche's anti-modernist and anti-capitalist philosophy. Ecology thus found an appropriate niche in a country where feudal structures were still strong, where national unification had become a driving objective, and where folkloric myths still retained power in the popular imagination.

The importance of ecofascism is not merely historical. On the one hand, some well-known ecologists in contemporary Germany have ressuscitated the old right-wing arguments, i.e. using ecology as a plank for nationalism and the rejection of multiculturalism; Janet Biehl's denunciation of Green activist Rudolf Bahro is particularly striking (Biehl and Staudenmaier 1995: 53–8). More broadly argued (i.e. beyond Nazi ideology and German neo-nationalism), if ecofascism is meant to designate any advocacy of "human cleansing" for ecological purposes, then it still holds sway currently in minority circles of the ecological movement. Ecofascism here merges with some trends within "radical ecology," which are actually much more reactionary than radical, and which remain quite inarticulate politically. We discuss this below.

Gaia and misanthropic ecology

Much of contemporary ecological thought either directly derives from or is, in some way, indebted to James Lovelock's research on the global ecology and his Gaia hypothesis (Lovelock 1979). As a biologist, Lovelock understands the principles of cyclicality and interdependence inherent in the working of any ecosystem, and so can defend their application at the global ecosystemic level. The argument would be difficult to refute in light of twentieth-century research on global ecology and, particularly, global climate, which has analyzed various types of interrelations between distant natural phenomena.

The Gaia hypothesis, however, takes the argument further by endowing the planet with a life that exceeds the sum of its component parts. This explains the use of the Greek term, referring to Earth as Mother. Gaia thus treats the planet as a single organism, adapting and surviving through its billion-year history, thus welcoming and shedding various forms of animal and plant life in the course of its evolution. As a scientific description of planetary survival through self-regulation, the Gaia hypothesis necessarily evaluates the contributions made by Earth's various species, and will corroborate what biologists understood a long time ago: that the key to ecosystemic survival is found at the bottom of the food chain, and that higher-level species are proportionally more "expendable" (even if their average death rate is small).

In sum, the political and ethical significance of the Gaia hypothesis lies in the language required for an assessment of Earth-as-living. The planet is a *bona fide* being; it may not think, but it lives. As a living being, it has needs that must be fulfilled and shows resilience in the face of assaults. Lovelock did not merely develop a heuristic device in the quest for environmental protection. Obviously, Gaia as heuristics is a powerful and helpful image, but taken literally, it will clash with ethics. As a scientist, Lovelock's intention was not to propagate naturalist doctrines that would decree the basic ecological irrelevance (or, indeed, danger) of the human species. Yet this is a lesson which ecological activists could presumably derive from Gaia, as they assess either the parasitic or, more bluntly, the "antibiotic" behavior of human beings through history. The Gaia hypothesis is thus crucial to misanthropic views that reduce the ecological crusade to the survival of "Mother Earth." The well-documented rift between "deep" and "social" approaches to ecology (which, we argue below, is not incurable) is reflective of this slippery alliance between naturalism and nihilism.

Deep ecology is a perspective emphasizing the mystical connection between humans and broader nature; however elusively, it suggests a profound symbiosis, a converging identity between human and animal, human and tree or human and mountain which clearly runs within Gaian principles. Extremist eco-activists with deep ecological affinities, such as found within the Californian group Earth First!, have uttered politically-laden statements which have hurt the ecological cause (B. Taylor 1991; Dobson 1995: 62). Using Gaian logic, one could indeed dismiss the human species as inherently unworthy of life on Earth, considering its location on the ecosystemic map and its tendency to upset natural equilibria, through presumably foolish attempts at uncovering the secrets of nature and harnessing its forces. From this perspective, epidemics and starvation in densely populated areas are welcome, while successful acts of "eco-terrorism" legitimately counteract the logger's or the miner's base assault on innocent nature. Whether the motivation is guilt or hate of the human species, the result is not merely the indictment of humanity (which, in and of itself, is defensible), but its draconian sentencing. Ecological thought becomes not merely an instrument of social resistance, but a tool of war – which is very different. The value of tolerance usually associated with ecology is absent here; compassion is arguably not a natural trait, and nature is not to be transcended.

Assessment

"Authoritarian ecology" is a generic expression which we use to indicate the possible relationship between saving nature and sacrificing freedom. In the long-run, that relationship may well be self-defeating. Radical ecologists would argue that nature and people are not ontologically separate, and so that the mistreatment of one is the mistreatment of the other. More precisely, they would argue, philosophies of domination open the door to the domination of living beings in all their forms, and so one cannot truly envisage an ecological society based on control or oppression; besides, oppression has historically produced resistance and revolution, which always entail some disorder likely to engulf surrounding nature. It is obvious that liberals favoring individual rights would be equally concerned with the consequences of an authoritarian ecology.

Nonetheless, eco-authoritarians obviously believe in their ecological credentials. It would seem sensible to argue, on the one hand, that if individuals or corporations cannot resist maximizing the short-term interest involved in anti-ecological behavior, then they must be disciplined and guided by some mortal God, to use Hobbes' expression. For students of politics, here is a tangible attempt at inserting ecological thought within political thought, with likely implications for IR theory. Enforcing conservation and preservation by the state entails legislation allowing a broad spectrum of "eco-friendly" interventions, from prosecution against industrial polluters to the creation of national parks.

However, invoking the modern state is fraught with complications, not the least of which is the assumption that the Green Leviathan knows how to define ecological sustainability. If that authority can ever be conjured, it may well collapse under the pressure of civil war. In practice, in a modern context, it will likely be co-opted by those it seeks to control (Finger and Kilcoyne, 1997) or be used, in all its renewed might, for repressive purposes that have much to do with state control and little to do with ecology. As Nancy Peluso argues, the state's "mandate to defend threatened resources and its monopolization of legitimate violence combine to facilitate state apparatus-building and social control" (Peluso 1993: 47).[12]

Alternatively, eco-authoritarianism may eschew the utilitarian underpinning of Green-Leviathan approaches, and see the repression of human freedom not merely as a means to ecological success, but as an end coincidental with ecological resurgence. This entails a strikingly different attitude vis-a-vis nature, a form of ecocentrism blending with mysticism. The term ecofascism seeks to designate part of this worldview, even if the term is derived from the name of an Italian party/regime not particularly known for its ecological prowess. In the German tradition, as a tool of a racist, authoritarian regime, it becomes central to the search for racial purity, with the glorious nation attuned to its roots in the land. Purged from its historical roots, ecofascism becomes a catch-all term for any attempt, state-based or not, at enforcing an ecological code of conduct regardless of human costs. The risks of repression inherent in holistic views of nature, such as Gaia, may not always be clear to the eco-neophyte in search of meaning in the midst of ecological crisis, yet they are real and are occasionally expressed by fringe activists.

Are there links already forged between this stream of ecological thought and IR theory? As we will see below, scarcity arguments, with their potential for international conflict and international management, are a mainstay of the "environmental IR literature." Meanwhile, the realist school of IR is arguably (and paradoxically?) imbued of the idealist current which so profoundly affected European relations in the nineteenth and twentieth centuries, and which, as we know now, is highly naturalist in tone; in other words, realists have analyzed the "reality" of an international political violence that partly emanated from nineteenth-century romantic calls for warfare, and which found particular expression in the blood-and-soil ideology of Nazism. In sum, if the language of ecology is one of control and violence, then it must have a bearing on IR theory.

RADICAL ECOLOGY

A radical philosophy seeks a profound revision of established social institutions. Radical thought, then, is usually political thought, whether the political component is explicit or not. From an ecological perspective, radical thought presumably commands a fundamental reassessment of established views of nature and of accepted institutions which pose a threat to nature's viability.

Stated as such, is there "a" radical ecological view? The prudent, and quite legitimate answer would be no. One could indeed refer to a series of well-known ecological schools which could all claim radical status: deep ecology, ecoanarchism, ecosocialism, bioregionalism, ecofeminism (among others). To complicate matters, the expression "social ecology" is often used as a synonym of ecoanarchism, while ecosocialism seems social enough to be fitted within social ecology. Meanwhile, bioregionalism apparently weaves itself into both deep ecology and social ecology, and ecofeminism is known for intense internal divisions which will puzzle any reader pursuing a neat classification.

We do not want to minimize the serious theoretical differences within the radical field. While academics and activists often defend their distinct contributions for narrow career purposes, the level of debate within radical circles is very high and is not merely centered on details. An ecologist reflecting on the control of women (ecofeminism) should well frown at the thought that all nature could be subsumed under an extended self (deep ecology), while a critic of capitalism who sees its destructive ecological impact (ecosocialism) could legitimately wonder whether a radical project should begin with such slippery language as "inherent value" (deep ecology, again). Alternatively, a theorist who situates ecological degradation squarely within long-standing, hierarchical patterns of authority (ecoanarchism) may react against perspectives openly sympathetic to the Gaia hypothesis and its reactionary overtones (deep ecology, bioregionalism).

However, we do not believe that the radical literature is exceedingly divergent, and will therefore stress some of its commonalities as we identify a radical response to utilitarian and authoritarian ecology. Among the five schools mentioned above, all but ecosocialism are arguably ecocentric, i.e. assuming that "nature" does not fundamentally exist to serve humans. Furthermore, all radical ecologists arguably seek to uncover the various processes leading to domination and control (of individuals and nature in the broad sense). Inspired in part from non-Western philosophies and practises (such as Taoism and First Nations cultures), extensively from Western critiques (Marxism, Critical Theory, feminism, postmodernism), and a direct rapport with (and scientific knowledge of) wild nature, radical ecology presumably seeks freedom for all life forms and peace for all human beings.

The defense of a "convergence argument" within radical ecology is controversial, and lies at the heart of a bitter debate between deep and social ecology, itself turned into a debate between supposed ecocentrists and anthropocentrists. Yet the division between deep and social ecologists is perhaps overstated. On the one hand, the recent intellectual "entente" between social ecology's most famous spokesperson, Murray Bookchin, and deep ecology's most famous activist, Dave Foreman, underscores the philosophical and practical links between "saving nature (for nature's sake)" and "freeing man/woman" (Bookchin and Foreman 1991). Several analysts have since given credence to the convergence argument. Michael Zimmerman, for instance, concludes that "despite their sometimes heated debates, deep ecology and social ecology have much in common": they both value nature intrinsically, reject a facile human-nature dichotomy, insist on nature's complexity and the need for wilderness preservation, and are very critical of the hierarchical, centralized and plundering character of the modern project (Zimmerman 1994: 151–2). Zimmerman also quotes Bill Devall, a central figure in deep ecology, who defends his adherence to social ecology in view of his own interest in (and denunciation of) capitalism as a cause of ecological degradation (Zimmerman 1994: 169). Likewise, the ecosocialist David Pepper acknowledges the essential coexistence of deep and social ecologists under the anarchist umbrella (Pepper 1993: 152). Even philosopher Bryan Norton, who is primarily concerned with the rapport between American conservationists and preservationists, appeals to a convergence argument linking, amongst others, deep ecology and ecofeminism (which has its own affinities with social ecology – Norton 1991: 197–8).

All radical ecologists, irrespective of their battlehorse (animals, the poor, humanity, women, forests, etc.), are concerned about ending organized, structural violence against exploited life forms; in most cases, this will specifically entail the search for a rekindled bond between human and nature; in all cases, but often implicitly, the goal of ecological peace/freedom will dictate a rejection of the conventional methods of "knowing." Radical ecology must be understood as an emancipatory critique of modernity. In this sense, much of radical ecology could be understood as "non-shallow ecology," an "ecology" that understands the fundamental incompatibility between growth-oriented, top-down ideologies of power and the type of order suggested by nature. All "non-shallow" ecologists, then, at the very least, would agree that life is not reducible to mechanical operations. Most such radicals would accept the postulate of nature's "intrinsic worth": the ethics of "what to save" may be debated, but they will agree that no forms of life should be exploited, and that any large-scale intervention in nature will trigger an ecological imbalance of potentially dangerous proportions. Radical ecology thus arguably manifests four sorts of concern: to present an alternative picture of nature (scientifically and ontologically); to reverse historical tendencies toward domination; to undermine the political project of mainstream science; and thus to uphold a frugal and egalitarian ethic that challenges that of modern life.

Selected sources of radical ecological thought

We will review in substantial detail below four of the five schools of radical ecology mentioned earlier; the status of bioregionalism is more tenuous, and so it will receive only passing attention. The remaining four all contribute (though in uneven ways) to a radical project of social reconstruction combining ecological living with social peace and freedom. Irrespective of the labels, these ecologists all understand that "saving nature" requires political and economic theory, as well as a reconsideration of the modernist ontology and epistemology linking humans to their "environment." Ecosocialists will be harder to defend from this perspective, but they still contribute positively to the radical critique.

At this juncture, it is important to recognize that the radical philosophy of ecology of the late twentieth century is derived from older sources and coexists with other radical traditions which may not specifically discuss "ecology." In other words, radical ecology is part of a general movement of dissent, marked by towering intellectual (and political) figures whose writings are not typically known as "ecological," yet whose ecological credentials seem unmistakable. So as to cast ecological thought in the broadest possible light, and so as to best explore the avenues through which IR theory may be subjected to an ecological critique, it seems worthwhile to address some key authors whose writings appear particularly relevant to the mission of radical ecology. We discuss five of them: Thoreau and Kropotkin from the nineteenth century; and Gandhi, Mumford and Schumacher from the twentieth. The list could be easily extended, yet we believe that this selection captures very well many of the key themes discussed by the contemporary schools of radical ecology.

The American transcendentalist Henry David Thoreau (1817–1862) is well known for his expressive naturalist writings, equally admired by social ecologists and less politicized preservationists. But Thoreau's significance as a radical political ecologist should not be de-emphasized, for his pamphlet on civil disobedience influenced not only the entire anarchist tradition, but also any thinking which postulates a relationship between the control of human beings and the control of nature. Thoreau, the eccentric recluse of Walden Pond, criticized what he understood as the alienating and destructive power of the state:

> I please myself with imagining a State at last which can afford to be just to all men, and to treat the individual with respect as a neighbor . . . A State which bore this kind of fruit . . . I have imagined, but not yet anywhere seen.
> (Thoreau 1962: 104)

"Ecology" becomes here a form of libertarianism. Thoreau asks: "Must the citizen ever for a moment resign his conscience to the legislator"? The response is scathing, a striking presage of Mumford: "The mass of men serve the state . . . not as men mainly, but as machines, with their bodies"(Thoreau 1962: 86–7). By "quietly declar[ing] war with the State" (ibid.: 100), Thoreau denounced its adventurous schemes, at home and abroad, for control and profit. His ideal is clearly that of a "local life," close to nature, and untrammeled by far-off commitments which can only serve a globalizing elite.

Contemporary radical ecologists have also found substantial inspiration in the works of Russian emigre Peter Kropotkin (1842–1921). As we will discuss below, Kropotkin's anarchist writings fundamentally influenced the ecoanarchist Bookchin, yet Kropotkin's ecological message arguably radiates beyond Bookchin and his devoted followers. The link between Thoreau and Kropotkin is apparent; while the former's anarchism is imputed, Kropotkin's is self-declared, following a long line of famous "anti-statists" (from Godwin to Bakunin), but articulating the anarchist ideal in an eminently ecological, peaceful manner. As with Thoreau, the biologist Kropotkin developed an ethic of non-violence based on his love for and knowledge of nature. The key to his thought is a particular reading of nature emphasizing mutual aid rather than ontological conflict (Kropotkin 1955 [1902]). In a crucial historical period (the late nineteenth century) where Darwin's research was increasingly appropriated by proponents of the "conflict model," Kropotkin sought to rescue Darwin's own insistence on the sociability of beings – their "natural preservation," a term which Darwin wished he had favored over "natural selection" (ibid.: 117). Kropotkin, through his many travels, read nature as a cooperative cycle of life, and extended his observations to the social world – where he could effectively document efforts by local groups at bypassing the state in furthering particular projects. Politically, the logical conclusion was to formally defend the system of anarchy, requiring cooperative (thus decisional) input from the grassroots in all social construction: "No ruling authorities, then. No government of man by man; no crystallization and immobility, but a continual evolution – such as we see in Nature. Free play for the individual, for the full development of his intellectual gifts" (ibid.: 59).

Kropotkin could not completely shun the modernist pressures of the time, and surely underestimated how new projects, stimulated by the new technologies, could actually void in the long run the ethic of cooperative peace that he embraced. Indeed, Kropotkin's language is often utilitarian (see ibid.: 55), while his emancipatory aim was still very much related to Marx's. But Kropotkin's contribution should not be de-emphasized on that account – and indeed Bookchin owes much to him. The logic of Kropotkin's argument and its naturalist base were leading to peace, even if the cosmopolitan Kropotkin did not completely share Thoreau's frugality and "locality." Mohandas K. Gandhi (1869–1948), on the other hand, stood very close to Thoreau on this latter point, and readily conceded his influence upon him. The Gandhian ethic parallels much of the contemporary discourse about positive peace and emancipatory ecology, including the advocacy of non-violent resistance to imperial aggressors, which defuses the spiralling cycle of destructive energy; the insistence on community life, grassroot involvement and basic-need production; and reliance on simple production techniques rather than sophisticated machines (Gandhi 1961).

Gandhi did not write political treatises per se, and while his wisdom literally laid bare the problems of modernity, radical ecologists did have to look elsewhere for theoretical guidance; moreover, his essentially patriarchal rapport with women has understandably bothered many feminists, even in the ecological tradition (McAllister [ed.] 1982). But the social ethic has stood there for many ecologists to vindicate. The most striking examples are in Schumacher's writings (see below) and in the work of Southern ecologists, particularly Vandana Shiva (1988, 1993) and Ramachandra Guha (1990). Yet Gandhi's presence is no less fundamental in deep ecological texts, attesting to his spiritual commitment to self-realization in a communal setting and, obviously, to his intellectual bond with Thoreau (Devall and Sessions 1985: 232; Naess 1989: 146–8).

In contrast to Gandhi, the self-taught American philosopher, Lewis Mumford (1895–1990), one of the most prolific ecologists-without-the-name of this century, has gone largely unnoticed by social and political theorists, within and outside of ecology.[13] However, his writings remain of utmost importance, both for the chosen theme (organum vs. machine) and for its bold, imaginative treatment. For instance, there is much of Mumford in Bookchin and in the ecofeminist Carolyn Merchant, although no real recognition of him. He did not specifically partake in ecological debates; his discussion of technics and urban design, however, yielded powerful ecological statements on peace. He deserves inordinate space here, both for what he said and for his otherwise mysterious absence.

Mumford's ecological statement can be gleaned from a review of his *Pentagon of Power*, the second part of his master treatise, *The Myth of the Machine* (1970). The basic message is not altogether original: overtaken by the mechanical model of Newtonian science, contemporary society has drifted toward a non-organic anti-culture of power, speed, and limitless pursuits. However, Mumford's momentous contribution stands elsewhere, in his characterization of mechanization as myth. The "myth" is not to be understood merely as illusion or falsity, though Mumford obviously agrees that the apparent achievements of mechanical society are essentially expressions of anti-life. The myth is, in fact, to be understood in the literal sense, as the construction of a cult, as a new religion – paradoxically, as the embodiment of genuine human feelings (fear of death, desire for power), and not as the pinnacle of rationality. The mechanical metaphor is indeed so pervasive that Mumford is able to describe society itself as a "megamachine": a gigantic operation, composed of human parts, destined to serve the gods of power.

Mumford comes to his conclusions through a very personal, and highly original, reading of history and philosophy. The parallels established between the "Pyramid Age" and the Enlightenment society are very suggestive – in both cases, instances of human entrapment in the pursuit of irrational heights through a formidable, technics-based harnessing of human reason. While many authors have analyzed the modern paradox pitting instrumental reasoning against irrational ends, Mumford clearly established its historical precedents. Ancient Egypt might have been governed by a powerful dictator, but the centralization of energy in the quest for seemingly absurd objectives is no less observable today; such energy is simply channeled through vested interests in the science-government-business triad.

In sum, Mumford denounces the thoroughly anti-organic make-up of a power society. The attack is not against science and technology per se, both of which can contribute to an ecological design of "plenitude." He does insist, however, on the doomed reductionism of mechanical thought, what amounts to a despiriting caricature of life:

> No machine ... can even theoretically be made to replicate a man, for in order to do so it would have to draw upon two or three billion years of diversified experience. This failure to recognize the importance of cosmic and organic history largely accounts for the imperious demands of our age, with its promise of instant solutions and instant transformations – which turn out too often to be instant destructions and exterminations.
> (Mumford 1970: 91)

The machine model, the eminent misreading of organicism, thus commands particular political structures and societal objectives which threaten social (and ecological) stability. Hierarchical (elitist) systems are devised so as to release energies in pursuit of quantitative utopia: more, better, faster. The centralization of power becomes a sine qua non for this anti-ecological project – be it in its blunt totalitarian form (bolshevism, fascism, corporatism) or in its subtler expression (capitalist technocracy). Rejecting the steady state, intolerant of cultural diversity (which slows the process of expansion),[14] the totalizing mega-machine is inherently geared to conquest. Warfare is merely the ugly culmination of this relentless drive toward change, toward the extraction of energy and the transformation of matter: "imperialism, which resulted in the temporary subjugation of the major territories of the planet by Western industrial and political enterprise, had its ideal counterpart in both science and technics" (Mumford 1970: 119).

Finally, in the work of British economist E. F. Schumacher, we find a contemporary statement of Gandhian ethics by a Christian scholar, and one of the most lucid testimonies to the virtue of an ecological society. Schumacher's celebrated book, *Small Is Beautiful* (1973), is a scathing attack on the common wisdom of liberal economics, correctly described as a profoundly anchored metaphysical creed (yet devoid of all spirituality), whose inherent logic leads to the destruction of the natural capital upon which the totalizing capitalist system is upheld. The famous essay (chapter 4) on "Buddhist economics" neatly summarizes Schumacher's blueprint for a better world: a world of humane proportions (a "globe of villages"), minimizing wants and consumption, using progressive technologies only[15] for the well-being of all members of the community.

The concept of peace is essential to Schumacher's ecological thought, as it is arguably to all radical ecology. Schumacher does not provide a precise definition of peace, yet few could misread his line of thought. In his critique of technology, he argues that peace is indivisible: "how then could peace be built on a foundation of reckless science and violent technology?" (Schumacher 1973: 34). Quoting Dorothy Sayers: "War is a judgment that overtakes societies when they have been living upon ideas that conflict too violently with the laws governing the universe" (ibid.: 38). Peace, then, is scarcely a function of appropriate power distributions, centralized leadership, or material growth. The ecological understanding of peace compels an examination of all forms of violence, locating their sources at various societal levels; war is but one expression of violence, and not a *sui generis* phenomenon. Ecological thought thus suggests that assaults on peace will necessarily flow from violations of those "natural laws" favoring permanence, and which are best captured by the question of size.

In emphasizing appropriate size, Schumacher argues that ideologies which favor or are conducive to large constructions necessarily entail disempowerment, marginalization, and impoverishment. Capitalism is of course at stake, for it is a totalizing economic system, dictated by greed and envy. As any totalizing device, it creates artificial (hence dangerous) distortions in a society, simplifying what should be complex and complicating what should be simple.[16] In our modern societies, survival appears contingent on forces totally out of one's control; securing basic needs inevitably requires violence, with obvious ecological implications (ibid.: 59).

It is important to recognize, in response to Bookchin's reading of Schumacher, that, on the one hand, small scale is not considered a sufficient condition for non-violence or non-repression. Smallness is part and parcel of a larger ecological philosophy that stresses ethics and metaphysics as much as physical nature. In typically Aristotelian fashion, the rational individual must be committed to a sense of higher purpose, to goodness and respect. Science can provide us today with a better understanding of the fragility of nature, but human beings must not be enslaved by instrumental reasoning and pursue harmony with nature:

> Our reason has become beclouded by an extraordinary, blind and unreasonable faith in a set of fantastic and life-destroying ideas inherited from the nineteenth century. It is the foremost task of our reason to recover a truer faith than that.
> (Schumacher 1973: 93)

Furthermore, the small community favored by Schumacher still allows for agglomerations of appreciable size, and does not seek to gloss over the dangers associated with rigid social hierarchies and deep-seated superstitions.

In sum, the purpose of the discussion above was to emphasize that contemporary arguments proffered by radical ecology are derived to a significant extent from a body of thought rarely described as "ecological." This is to show how radical ecology, in its attempt to free man/woman/nature from structures of domination and exploitation, is a wide-ranging philosophy. It is a critical reflection on modernity which begins with idealist, materialist and romantic currents of the nineteenth century, and which is still learning from post-war reactions to the dangers of bureaucratization and instrumental reason; Marx, Weber, Marcuse and Foucault all seem essential in formulating (and appreciating) an incisive, radical ecological critique.

Deep ecology

The inclusion of deep ecology within the radical stream may be controversial, for it has been accused of flirting with mysticism and authoritarianism and ignoring the complex, social bases of ecological degradation. We believe nonetheless that its presence here may be defended, and that deep ecology should not necessarily be construed as a reactionary and/or a politically naive response to the ecological crisis.

Deep ecology is a term coined relatively recently by Norwegian philosopher Arne Naess (1972), and used to describe a set of ecological precepts that would oppose the "shallow," anthropocentric ecology upheld by the utilitarians discussed earlier in this chapter. Central to deep ecology is the principle of biospheric egalitarianism: in this ontology, no one is given legal or moral dominion over the rest of nature; this reflects a reading of nature as organically integrated and surviving through cyclical processes of creation and re-creation. Gaian traces are more than visible within this stream, whose sources are found in some of the nineteenth-century movements briefly mentioned above. More precisely, deep ecologists are particularly influenced by some of the great "preservationist" activists of North America and Australia, new worlds of bountiful forested land threatened by run-away industrialization. Preservationism is best understood as a movement designed to fence in large tracts of land and protect biospecies (Norton 1991). While the motivation for preservation could be purely utilitarian (e.g. for upper-class leisure), in this case it is better understood as a romantic pursuit, as a means to secure the necessary room for the individual's contemplation of, and harmony with, nature within a context of austerity. Deep ecology thus may well appear as a form of spiritual renewal bordering on paganism, and so one may understand why its critics see it as incompatible with a rational, radical movement for political change.

Still, as a holistic philosophy, deep ecology is obviously indebted to various scientists, particularly biologists, who have stressed the interconnected character of nature. One may legitimately see the beginning of this scientific tradition in Scandinavia, with the reputed eighteenth-century Swedish biologist, Linnaeus (Worster 1985). More recently, V. I. Vernadsky, who suggested the concept of the noosphere, wrote persuasively that "no living organism exists on Earth in a state of freedom;" "all organisms are connected indissolubly and uninterruptedly" (Vernadsky 1945).

This said, deep ecology is not merely a European tradition, for it would hardly have evolved without the contribution of illustrious American naturalists, and, lately, Australian philosophers and eco-activists. In the United States, much has been written on the role of John Muir in shaping modern American environmentalism. Muir built on foundations laid by Thoreau. A biologist and a social activist (yet a loner much more at ease in the wild), Muir fought successfully for the creation of national parks in the nineteenth century and established the now world-renowned Sierra Club. Muir's radical mission was in contrast to Gifford Pinchot's conservationism. Rebelling against his harsh Christian upbringing and against the ideology of production which he saw linked to Christianity's ontological dualism, Muir sought meaning in a direct contact with wilderness that is denied by modern, urban society (Eckersley 1988). The contemporary Australian deep ecologist, Warwick Fox, writes of nature as the human being's extended self, or perhaps more correctly, he views nature as a series of merging selves that include the human's (Fox 1990), and this is an ontology which Muir would share.

Muir's intellectual successor, Aldo Leopold, would clarify the early deep-ecological quest for ecological harmony by invoking the idea of a land ethic, an idea with obvious political implications. It was Leopold who poetically wrote about the importance of "thinking like a mountain," of recognizing that "a thing is right when it tends to preserve the integrity, stability, and beauty of the biotic community" (Leopold 1949: 262). Leopold thus stated a point that was at times implicit in the nineteenth century, but that became explicit in the writings of contemporary deep ecologists: social instability (violence, undue suffering) is rooted in human violence against the land. Since the land ethic was enunciated, a vague statement with seemingly little political guidance has developed into a contemplative, deep ecology that is far from insensitive to the political and economic sources of ecological degradation. Naess' work is more than suggestive in this regard: heavily influenced by the Gandhian ethic of nonviolence and economic simplicity, Naess understands that "power analysis is necessary" and that the long-term peaceful future of the planet is inescapably tied to autonomous, non-violent struggle against oppression (Naess 1989: 131, 148).

In sum, if deep ecologists, for obvious reasons, have been wary of modernity, there is actually little in their writings that would suggest an indifferent acceptance of ecological "barbarism," i.e. an ecology indifferent to humanity. As Naess writes:

> The principle of biospheric egalitarianism defined in terms of equal right, has sometimes been misunderstood as meaning that human needs should never have priority over non-human needs. But this is never intended. In practice, we have for instance greater obligation to that which is nearer to us. This implies duties which sometimes involve killing or injuring non-humans.
>
> (Naess 1989: 170)

Admittedly, if deep ecology is reduced to an elusive metaphysic, then its identity as a radical political movement is put into serious doubt (Tobias 1994). If deep ecology is mostly remembered for its insistence on developing a purely transcendental communion with nature, then it loses any status as a political theory open to critique. The oft-quoted Earth First! activist, Christopher Manes, says as much here:

> The soundness of these ideas cannot be ascertained by philosophical analysis so much by the role they are playing in a culture facing a period of ecological upheaval. . . . This kinetic aspect of radical environmentalism has been lost on many commentators, who understand this new cultural force as a body of ideas rather than a body in motion.
>
> (Manes 1990: 21–2)

However, it would seem unfair to deny deep ecology all claims to a political theory designed to relieve nature from various networks of control and abuse. In this sense, deep ecology is much closer to Bookchin's ecoanarchism (or "social ecology") than the latter would have once admitted. Naess has clearly stated that "supporters of the deep ecology movement seem to move more and more in the direction of nonviolent anarchism," which is a concession to Bookchin (Naess 1989: 156). Bill Devall and George Sessions, also key figures of the movement, have praised Bookchin in their landmark book, and offered an anti-dominant "worldview" combining references to both "intrinsic worth" and appropriate husbandry of nature (Devall and Sessions 1985: 69); thus they hoped to demonstrate that deep ecology is not inimical with reason, that deep ecology is compatible with political and economic projects of "permanence" (to employ Schumacher's term).

Social ecology

The expression "social ecology" was coined by Bookchin, so as to label his ecological philosophy heavily based on anarchism. Social ecology is meant as a counterpart to deep ecology; it suggests that deep ecologists do not sufficiently engage in social theory, and that ecological degradation is best understood as the product of relationships of domination and exploitation.

Before moving any further, we should certainly clarify some basic points, so as not to be confused with various terms: a) "social ecology" and "ecoanarchism" are often used interchangeably, since both labels are connected to Bookchin; b) it would not be inappropriate to encompass within social ecology other ecological perspectives which do account for the social sources of ecological crisis; ecosocialists are presumably "social ecologists," and so are bioregionalists; however, we will follow established practise by considering ecosocialism separately; c) many ecofeminists are also serious students of social ecology (some are inspired by Bookchin), but we will treat them separately as well in view of their specific commitment to the cause of women (yet understanding the point that all radical ecologists must, by definition, endorse the emancipation of women); d) while social ecology is meant to oppose deep ecology, we recall that this opposition is overstated; deep ecology is indeed guided by the same anarchist principles upheld by Bookchin, even if, admittedly, deep-ecological writings by no means approach the historical and philosophical depth displayed by Bookchin (and other "social ecologists").

Bookchin has purged his anarchism from some of its nineteenth-century confidence in scientific and technological progress, embracing the Critical Theory of the Frankfurt School in emphasizing the limits of instrumental rationality. Yet he has carried his own critique further down the radical path by advocating the outright abolition of all structures of domination, particularly the state. For

Bookchin and others such as George Woodcock (1992), truly ecological thought is necessarily anarchist thought.[17] Environmental destruction is a function of social hierarchy; while capitalism represents the pinnacle of control and exploitation (by exploiting the forces of science and technology), it remains one instance of a series of hierarchical patterns that have characterized human history. Bookchin's ecological thought is best understood as a broader reflection on (and pursuit of) freedom. Most societies in history have "suffered" from hierarchical control; as Marx wrote, the history of civilization is largely an attempt by elites to control the labor of people, i.e. to extract from "nature" some use-value leading to capital accumulation. Hierarchies, in their essence, repress freedom and make use of life as resources. Thus the logical proposition is to sever the chains of domination, so as to establish truly democratic communities – communities of empowered individuals who will obviously have their own interests at stake, who will not be manipulated by authority, and who will rekindle their natural cooperative links which they were forced to shed hitherto.

Bookchin's ecoanarchism derives from several philosophical sources, almost all of which contain at least some ecological dimensions. Among these (and beyond Kropotkin) we may include the Greeks, Rousseau, and Mill. Rousseauian themes are particularly striking, for Rousseau spoke vociferously about the kind of direct democracy espoused by Bookchin. Admittedly, Rousseau's social contract hinged upon the creation of a modern state, yet we may wonder how different the Rousseauian contract is from Bookchin's municipal confederalism. Rousseau's own reading of the Greeks, along with his fondness for the peasant community of Switzerland, shaped his commitment to an organic community that would involve all able citizens in the decision-making process; direct democracy would require a simple setting, and so Rousseau spoke affectionately of agricultural communities yet untainted by the corrupting forces of science, elitist art and capitalism (supported by the state).

Undoubtedly, the libertarian Bookchin would be wary of attempts at establishing a "general will" that would repress the necessary diversity of an ecological society, and so here, Bookchin flashes liberal roots that are particularly evocative of Mill. Yet the ecological society is meant to be organic: organicism is a very positive word for a radical ecologist, and if one recognizes that Rousseau understood the general will as emanating from society, then Bookchin's affinity with him becomes apparent.

As an anarchist, as a Kropotkinian, Bookchin assumes that human beings are essentially cooperative, and that conflict and egoism are social products; the natural "self-interest" postulated by classical liberals (Locke, Smith – and Hobbes) was in fact apparently ordered by bourgeois society (and, historically, by all hierarchical patterns). Bookchin echoes Rousseau once again. An ecological society requires the recovery of this "true" nature – a nature documented by Kropotkin's extensive travels and wilderness studies. Evidence from nature suggests that life is not a survival of the fittest, but a cooperative cycle of biotic enrichment – survival of the most ingenous in a fully interdependent setting. Thus, a radical ecology requires the expurgation of authority structures that turn human beings against one another, abolish their freedom, and instill false needs which the natural environment cannot bear.

In sum, by linking the health and freedom of nature with the democratic control of political communities, Bookchin adapts established currents of philosophy to the modern context – and Bookchin is a modernist, for he is committed to salvaging reason from what he considers to be the disastrous collapse of the Enlightenment project. This thesis is also defended by Tim Hayward (1994), who defends the continuing validity of the Enlightenment's emancipatory ideal. Bookchin is opposed to a pagan, mythical, reactionary ecology that, most probably, would jeopardize freedom. By insisting on creative reasoning, as opposed to a mere instrumental reasoning serving utilitarian gods, Bookchin shows his fascination with the Greek polis and vindicates Mill's own qualification of Bentham. Of course, Bookchin would reject Mill's commitment to representative democracy and continued adherence to private property, yet he could not deny Mill's attempt at favoring individual freedom through self-development, i.e. through cultural development in a context of free and honest intellectual exchange: "[we] must count on the probability that normal people have the untapped power to reason on a level that does not differ from that of humanity's most brilliant individuals" (Bookchin 1989: 198).

In the midst of his apparent reconciliation with deep ecologists, Bookchin wrote: "One of my goals is to foster the development of a non-hierarchical ethics of complementarity among humans *and* between humanity and non-human life. This should be the fundamental starting point . . . of the radical ecology movement" (Bookchin and Foreman 1991: 133). Whether the ecoanarchist really has buried the hatchet remains unclear, but the quote is nonetheless indicative of some consensus within radical ecology. As a social theorist, Bookchin's normative goal is a libertarian form of freedom, one that cannot exist if nature is commodified and controlled, as it is today by an economic elite. Yet this argument goes beyond that of ecosocialism, for ecoanarchism's ontology unites humans with other life forms and thus postulates that any relationship of domination entails, in and of itself, a control over a life form whose nature is to be free.

Furthermore, as a social theorist, Bookchin reflects on the proper setting for sociality, and this explains the substantial attention given to the city as an ecological setting (Bookchin 1992). This does contrast ecoanarchism with deep ecology's sustained commitment to wilderness, yet this is not to suggest – quite the contrary – that Bookchin's "nature" is reducible to urban gardens. The point, rather, is that an ecological society is quite compatible with an urban environment of modest proportions that stimulates the aesthetic and creative drives of humankind, while ensuring some level of material comfort through the use of "alternative" technologies.

To conclude, it would be useful to indicate that the ecoanarchist character of Bookchin's social ecology reverberates in the bioregionalist approach popularized by American journalist Kirkpatrick Sale (1996; also Snyder 1990). Bioregionalism is actually an interesting synthesis of deep and social ecology, as Sale recognizes the value of Gaia and its compatibility with a decentralized political and economic program. Rejecting the concept of "ecosystem" as reductionist and utilitarian (i.e. as the product of a scientific mind-set designed to understand and presumably control the mechanics of nature), Sale substitutes the concept of "bioregion," a rather loose term identifying the natural borders of a community or a group of communities. By living within bioregions, human beings are able to achieve the necessary harmony with nature which should characterize the ecological community. In other words, Sale's call for decentralization and appropriate size is quite reminiscent of Bookchin, Schumacher and others associated with the radical tradition. Ecology is surely political theory here, as Sale condemns the historical process by which private property replaced the commons, and by which the impersonal urban environment (guided by the profit motive) broke the links of community interdependence and replaced them with a fundamental dependence between consumer and supplier (Sale 1996).

The status of ecosocialism

Should ecosocialism belong in a review of radical ecological thought ? We can answer yes, considering that Marx remains inextricably linked to this literature (even if ecosocialism and eco-Marxism should not be conflated), and that Marx is obviously a cornerstone of radical thought. On the other hand, can we label "radical" an ecological school whose anthropocentric arguments are frequent? However, if we imply that radical ecology is synonymous with ecocentrism, then we are uniting "green socialists" with utilitarians and assuming that both are committed to an "industrialism" that will ransack the planet, irrespective of property relations. In the process, we may be committing the fallacy of labelling "socialist" Soviet or Chinese experiments that have arguably very little to do with Marx. On that account, ecosocialists will defend their vision of a radically altered society, one that will *necessarily* produce a harmonious relationship between humans and other life forms.

Questions about the ecological status of ecosocialism inevitably rise in view of Marx's own commitment to the forces of progress, i.e. science and technology (Eckersley 1988). Such optimism was the norm in the nineteenth century; as we saw above, even the anarchist Kropotkin succumbed to it. As an economic historian, Marx obviously recognized the possibility of scarcity, but understood it as the result of capitalism's inefficiency; egalitarian relations of production could presumably reestablish a proper management of natural resources (Benton 1989; Carpenter 1997). The influence of liberal economic theory on Marx was most evident in his labor theory of value and his concurrent commitment to an improved standard of living for the masses. Echoing Locke, and more immediately drawing on Ricardo, Marx argued that natural resources had no inherent value, and thus became valuable only when mixed with labor, i.e. transformed by the rational human being in pursuit of some project. Of course, Marx opposed the liberal argument that resource use justified (naturally) the establishment of property rights; but the key here is the anthropocentric view of nature, which is not compromised.

Marx, the materialist, very much appreciated the ingenuity of the liberal order, and could logically argue that such productive capacities would be retained under communism. Marx, the humanist, sought the dignity of the human being (violated by capitalist forces of alienation and exploitation), and would reject any attempts at mythicizing nature or at using the agricultural community as a means to pacify the revolutionary ardor of oppressed groups. And so Marx, proponent of *praxis*, claiming that teleology still requires action, emphasized the historical role of urban movements and showed predictable contempt for that wide majority whose livelihood was inextricably bound to the land.

For ecosocialists of the late twentieth century, social justice lies at the core of their political theory — not libertarian freedom or Gaia. In typical Marxist fashion, healthy nature is an appendage to the equality of social relations of production. Mastering nature is acceptable, as long as its benefits are not confined to a minority (Leiss 1972: 197–8); and, as Pepper (1993: 3) mentions, "we should proceed to ecology from social justice and not the other way around."

Is this a marginalization of ecology in a school of ecological thought? This would seem harsh. Ecosocialists see no reason in belittling their ecological identity on account of anthropocentrism. They are clearly appalled by modernity's relentless attack on nature (Lipietz 1995), as much as economic liberals who genuinely believe in market approaches to environmental management. Ecosocialists will argue, however, that an ecological society requires planning. It will not come about magically through romantic value shifts prompted by wilderness living or spontaneously, from below, in an anarchistic society. As a contemporary reflection on the violence of capitalism, ecosocialism recognizes what exploitative relations of production can mean for nature (writ-large), and argues that peace with nature necessarily requires the end of capitalism. Domination, in and of itself, is not the ecological demon perceived by eco-anarchists; one could argue that ecosocialists know relatively little about the history of domination outside of capitalism, yet it is still obvious that the modern quest for profit and value-extraction has produced an assault on nature of such historical proportions that even the "limited" ecosocialist agenda is defensible from the perspective of a radical ecology committed to respect of non-human life forms.

As mentioned above, the ecosocialist solution to the ecological crisis demands planning, what, in an interesting twist, *both* ecosocialist David Pepper and deep ecologist Robyn Eckersley define as an "enabling state" (Eckersley 1992: 175; Pepper 1993: 146, 233; also Gorz 1980). Invoking the state as a temporary measure is a classic Marxist argument, prompting the historic rift between followers of Marx and those of the anarchist Bakunin and still criticized today by many radical ecologists (Bookchin 1980: 287ff). Experiences with twentieth-century "communism" indicate the danger of statist solutions, yet we should remember that both Soviet and Chinese revolutions were peasant-led and overthrew essentially feudal (not capitalist) systems. A socialist solution to the ecological crisis could thus legitimately allow for some guiding force, stemming from above, and responsible for the transition from a propertied economy (which harms nature) to an economy stressing community "needs." Both (eco)anarchists and (eco)socialists seek the empowerment of the average citizen, yet socialists find the idea of spontaneous organization difficult to fathom.

In sum, ecosocialism is best understood as an approach to the ecological society based on traditional Marxist themes and socialist activism. The approach is statist to a significant extent, yet this does not preclude the effective mobilization of key groups, particularly labor; most environmental groups are actually seen as "bourgeois" (Pepper 1993: 247). Ecosocialism's radical character will be questioned, yet the approach arguably wishes for changes that would not be disputed by many eco-radicals: needs-based production, equality of status, appropriate technology, democratic decision-making, humane-size cities. This said, ecosocialism, as a guide for political action, is bound to remain torn between its "red" and "green" components. The split within the German Green Party provides us with a lesson in that regard; the debate centered on the possibility of embracing the concept of sustained growth, dividing a party which had derived support from workers and unions (which approved of growth) and more radical ecologists (who did not).

Ecofeminism

As the term indicates, ecofeminism seeks to combine, in an intellectual framework and in a political movement, two sets of concerns: the defense of women and the defense of nature. The basic argument is strong enough to qualify as radical: patriarchal society has sought historically to control all "subordinate" life forms, including women and broader nature; the oppression of women is, therefore, part and parcel of the ecological crisis, as women are subjected to the same dynamics by which value is extracted from "passive" nature.

Is this twin aggression merely reflective of the modern project based on Cartesian dualism and modern science and actualized in global capitalism (Merchant 1980), or does it have deeper historical roots – way back to Plato (Plumwood 1986, 1993)? As men have tended to equate womanhood with bountiful nature, is this an image that should be praised (as emphasizing women's nurturing and caring "nature") or rejected (as depicting women as passive, unable to think rationally, lead communities and engineer projects)? As much as we had suggested above a working definition of ecofeminism, we are quick to recognize some of the serious divisions within the field.

To some extent, we may be perplexed upon reading some reviews of the ecofeminist literature, which seem to identify the same internal divisions characterizing feminism as a whole. Hence, while one may understand that feminism encompasses both the left and the right, one may be surprised to read that "radicalism" and "ecofeminism" do not necessarily belong in the same sentence. Consider, for instance, Melody Hessing's three-fold categorization. She initially identifies a "cultural" ecofeminism which she also labels "radical"; here, "[w]omen's affinity with nature is celebrated by the glorification of natural processes – fertility, pregnancy and birth – and a celebration of their embodiment in women . . . nature is mother; woman is wilderness; women, like the land, are abused, violated, scarred." "Liberal" ecofeminism, on the other hand, strives for "social change within the existing socio-economic structure" while "socialist" ecofeminists "address this structure as the ultimate source of both women's oppression and environmental degradation" (Hessing 1993: 16–17).

Thus, is "ecofeminism" able to provide any consensual guide for the creation of an ecological society, based on some agreeable understanding of the forces leading to the twin marginalization of women and "nature"? The answer is not necessarily forthcoming. Will reason save women and nature, in creating an ecologically respectful society based on ecoanarchist principles (Biehl 1991)? In contrast, should women cultivate some mythical rapport with nature and encourage this communion through various rituals and practises of witchcraft? Most ecofeminists would resist the latter extreme. While Merchant (1980), echoing Mumford to some extent, tremendously helped the cause of ecology by explaining how modern science constructed a mechanical model of nature, there is always the tendency to conflate its opposite, namely organicism, into a rigid mold dominated by elusive myths and demanding unquestioned discipline from members (Biehl 1991: 98–90). If ecofeminism is to be a step towards the emancipation of women, then there are presumable risks in linking it with some of the least social branches of deep ecology; Plumwood (1993: 165ff) is particularly wary of Fox's image of nature as an extended self – this is ecology as the denial of difference.

From our perspective as students of international or global politics, we must particularly appreciate how ecofeminists have analyzed contemporary power relations and explained how women's contact with the land (in the South) can set the appropriate setting for a progressive challenge against the global power structure. The writings of Vandana Shiva have been discussed extensively in this respect. She has demonstrated several times how the global, capitalist forces of production have done violence to the land and thus threatened the overwhelming majority of Southern women whose livelihood is tied to it (Shiva 1988, 1989); as a supporter of the reputed Chipko movement in India, she has also indicated how a movement of resistance may be undertaken peacefully and successfully by the apparent weak. More broadly, ecofeminist writings have questioned the epistemological foundation of modern society, i.e. the dualist and mechanical approach to nature breeding a largely male scientific community, and perpetuating deeply rooted patriarchal biases through the "objectivity" of the detached mind (Easlea, 1981).

Assessment

We have included within the rubric of radical ecology various schools of thought that are not compatible on all points. We have emphasized a fundamental convergence between deep ecology and social ecology, well aware of Bookchin's regular attacks on deep ecologists (amidst attempts at bridging differences). We have discussed the radical character of ecosocialism, again well aware of its fundamental anthropocentrism which could liken it to mainstream environmentalism. We have broached ecofeminism, yet indicated that it may simply (and more subtly) restate feminist debates in which very liberal thinkers participate.

In this morass, we may still distinguish a radical contour that does encompass most of the authors to be associated with these schools. The radical view stresses that the ecological crisis is a crisis of domination and exploitation, and that a healthy planet Earth will require the abandonment of political and economic structures conducive to the commodification and enslavement of life forms. This is not a system that can be "fixed" through technocratic ingenuity. Radical ecology is a statement on the objectification of life, and so it stands just as much as an epistemological critique – a critique of dualism and mechanicism. The ecological society would require a decentralization of structures, although how much is not clear. It would require tempering the bureaucratic reliance on instrumental rationality, although how much "reason" should be sacrificed is to be debated. It would require an empathy with fellow life forms, although a Gaian ecocentrism could be as dangerous to human beings as an excessive (anthropocentric) denial of an "ecological ethic."

Without pretending to neatly integrate these radical schools, we do believe that they offer powerful ideas in our assessment of the "ecology of IR theory." As IR theory reflects on the conditions of peace, security, order and freedom, there is much within radical ecology that would allow for a reconsideration of the descriptive and prescriptive make-up of IR theory.

CONCLUSION

This book assumes that the interdisciplinarity characterizing contemporary studies of the social world can encompass two fields usually seen as mutually isolated, namely ecological thought and international relations theory. In other words, we argue that the two fields of study have overlapping problematiques. Ecological thought is (to a large extent) political thought, and so is IR theory. The two fields are united in their pursuit of some of the perennial values of political life, such as the ones mentioned a few lines above: peace, order, security and freedom.

While Ecology and IR are presumably related, there is (paradoxically) relatively little discussion of international political dynamics within ecological thought, and, similarly, little recognition within IR theory of the complex process by which ecological degradation occurs and how it may affect global relations. Our first major step, therefore, was to provide an overview of the major schools in ecological thought. We needed to probe into the philosophical baggage of ecological thought, to determine what political recommendations may emanate from the various assessments of "nature" and of the assault on nature. Let us keep in mind that naturalist assessments are as old as philosophy; on the other hand, the "death of nature" is a new development, raising (usually) critical reflections among ecologists about the forces of modernity (science, technology, capital accumulation, power concentration) and their impact.

In the process, we wanted to know if, for instance, Ecology could provide a radical language that could alter the usual conceptions of peace/order/etc. found in IR theory, or, to the contrary, if Ecology is just a rephrasing of common political themes, such as scarcity and interdependence. The skeptic could argue that ecological thought is not original thought, i.e. that it is as conservative or progressive as the philosophical tradition on which its various components are based. On the other hand, while we will easily grant the point that ecological thought indeed echoes many themes found anywhere from Hobbes to Marx, it is worthwhile to realize that the ecological crisis triggered to a significant extent a renewed interest in philosophical classics among contemporary students of politics.

Our review yielded a three-fold classification of ecological schools: a utilitarian branch, treating nature as raw material to be wisely used and properly managed within a liberal framework; an authoritarian branch split between utilitarian and neo-romantic offshoots, either advocating centralized solutions to environmental problems or (in a different exercise) seeking harmony in nature through violence; a radical branch, locating the sources of ecological crisis within established patterns of authority and exploitation, and urging a fundamental decentralization of political power and reconsideration of the modern project of growth and bio-control.

We can conclude that the utilitarian perspective (including that of the Green Leviathan) can be (and has been) rather easily encompassed within mainstream IR theory, while the radical approach may perhaps be incorporated with (or added to) the current critical attack on mainstream IR. Ecological thought has arguably succeeded in lending credibility to a description of nature stressing: a) its finiteness (i.e. there are limits to growth); b) its wholeness (causality is not a linear process); c) its diversity (biosurvival requires an elaborate "safety net"); d) its very long evolution (nature cannot be reproduced by a machine). These are points on which all ecologists would agree, yet not all of them will deduce a radical program from such observations. Arguably, the more interesting avenues for a critical assessment of IR theory are found in the radical school, since radical ecologists demonstrate how the "common good" is compromised by existing global institutions, including that of reductionist science – and so the radical ecological critique encompasses ethics, politics, economics and epistemology.

In the next few chapters, we will assess the extent of "ecological awareness" in IR theory, and examine how the normative content of IR theory may be reinterpreted in light of ecological thought. It may well be, as Ronnie Lipschutz and Ken Conca have indicated, that "most students of international relations have treated environmental issues as lower than low politics" (1993: 14n8). If that is the case, perhaps IR theory has not realized how ecological thought is arguably the very matter of political theory, that reflections on "nature" are so basic to governance (national or international) that they tend to be assumed rather than articulated, and that only an ecological crisis may bring that realization to the fore. Conversely, in a globalized world, ecological thought must account for political relationships as they unfold on a large scale; the "international" dimension of ecopolitical thought is rarely discussed specifically, and although much of the global *problématique* is arguably implicit within Ecology, it remains necessary for ecologists to tackle theories of international relations directly.

Chapter 3
Realism and ecology

INTRODUCTION

The next three chapters revisit IR theory from an ecological perspective. As indicated earlier, the exercise can be neither comprehensive nor definitive, considering the breadth of the literature and its many possible interpretations. We introduced "classical" and "structural" realism in Chapter 1. Here, we focus most explicitly on the philosophical roots of realism, and as such do not treat at great length the geopolitical tradition, concurring with Daniel Deudney, who asserts in an innovative article that "most of the contemporary public discourse of 'geopolitics' is more a thematic and rhetorical dimension of American state-centered realism and strategic studies than a distinctive or articulated theory" (Deudney 1997: 97).

The significance of the realist school appears self-evident, for it has done more than any other discourse to define international politics in terms of a logic of conflict between power-hungry statesmen and generals in the absence of authoritative international governmental structure. While a quick glance at the most widely read journals in the field today suggests that realism (or its multiple variants) is still a fixture, it is equally obvious that it shares the spotlight with theories from liberal and, even, critical perspectives. Nonetheless, we can argue that realism has provided the foundation of the formal discipline of IR, one that derives principally from a select group of American and European authors.

How should one approach this interdisciplinary exercise? For our immediate purposes, the pivotal question surely pertains to the ecological credentials of the realist paradigm. If realism has provided the ontological base for international studies, this leads to questions about whether it is at least partly responsible for the current ecological crisis, or, alternatively, is conducive to the creation of an ecological society. Answering this question will allow us, in the process, to ask whether there are readily apparent ecological themes within realist thought, and whether there may be such a thing as a "realist ecologist." This would entail moving beyond the conceptual confines of geopolitical analysis.

The answer to this may not be so clear once we shed caricature and give realism its due process. There is a tendency to dismiss any perspective that can be construed as maintenance-oriented (as opposed to transformation-oriented) as counter-productive. Most radical ecologists, described in the previous chapter, would argue that centuries of realpolitik have produced world wars and favored the development of military technologies whose ecological impact has been devastating. Others may point out, however, that realism includes a Hobbesian logic of national governance that may effectively address environmental problems through centralizing formulas, or at least inform the debates over collective responsibility with an unfortunately necessary pragmatic tone. Still others would contend that realist politics is prudent politics, and that prudence is a necessary ecological virtue.

In order to examine the congruence between realism and ecology, if it exists at all, we need an appraisal of key realist assumptions and recommendations, and hence of their likely implications for the state of the world. Those assumptions and their implications are not necessarily articulated in ecological language, and so the ecological baggage of realism may not be readily apparent. We rely on several landmark authors and texts in assessing the tradition. While readers would surely wish to extend or otherwise modify our list, we believe that key realist precepts may be identified, at the very least, from the works of E. H. Carr, Reinhold Niebuhr, Hans Morgenthau, John Herz, Raymond Aron, Hedley Bull, and Kenneth Waltz; we also include references to one German historian of the nineteenth century, Heinrich von Treitschke, whose realism was not that of the Cold War defense bureaucrat, but that of the nationalist idealist who inspired the twentieth-century realist reading of world politics. We also acknowledge (and briefly discuss) the philosophical roots of realism in Thucydides, Machiavelli and Hobbes.

Second, we need to look at the directions realist thought has taken since the advent of widespread ecological awareness (or, more precisely, awareness of the magnitude of environmental problems). Again, when this awareness begins is a difficult methodological question, but we can place it as part of the driving force behind the rise of key non-realist perspectives stressing interdependence in the 1970s. Robert Gilpin and others have suggested that the need for collective solutions has not changed the fundamental jungle mentality of IR today, and that the real questions to be asked by scholars remain linked to power distribution between states. Others, such as Joseph Grieco, stress the inherent difficulties of international organization and regime formation. Some, such as Thomas Homer-Dixon, reject the realist paradigm explicitly but, arguably, end up reinforcing it with an emphasis on state stability in the midst of severe environmental degradation. To some, the rise of environmental concerns as high-profile, and the diplomatic activity surrounding their discussion and partial regulation, may qualify them as part of the realist international agenda in and of itself. However, this would be a superficial employment of both global environmental problems and the realist perspective.

Although we might be surprised to find that realism is not without ecological tendencies and implications, it is less surprising to note that realism demands severe limitations be imposed on thinking about global transformation. In this sense it can hardly be expected to provide a firm or even loose foundation for the marriage of IR theory and ecology sought in this text. The realist understandings of change cannot travel far beyond interstate relations and subsequent redistributions of power. A realist framework would reveal evidence of ecological crises in related intergovernmental bargaining, yet the internationalization of production and the transnational suffering of impacted groups is far more pervasive. Nonetheless, the significance of realism becomes more pronounced when we consider its linkages with the conservationist and authoritarian schools of thought outlined in the previous chapter.

EVOLUTION OF THE REALIST WORLDVIEW

Although it is often dismissed as overly simplistic by critics, realism derives from complex roots. Some would identify it by way of Robert Keohane and Joseph Nye's distinction between realism and complex interdependence: realism is essentially preoccupied with (unified) state actors in international politics and with military security issues (Keohane and Nye 1977). It privileges interstate relations as the basic empirical referent for the field, and assumes that military confrontation is the most significant outcome. While this depiction of realism is not inaccurate, it exaggerates the divisions within IR theory while omitting necessary references to the origins of realism. Many are left with the distinct impression that realism is purely a recent academic construction.[1] In fact, realism can only be understood historically and philosophically; and, while it is often depicted as an American foreign policy doctrine of sorts, it was "in large measure a transplant from continental Europe" (Kahler 1997: 25).

The deepest roots of realism may be found in writings by Thucydides, Machiavelli, and Hobbes (although the latter two wrote political theories whose breadth vastly extends beyond the narrow reaches of realist thought).[2] In the *Melian Dialogue*, Thucydides showed a "reality" of politics that ordered the destruction of enemies, however defenseless, so as to forestall any future rebellion against a conqueror. His account of the Peloponnesian War has served as the basis for realism's stress on interstate relations and the balance-of-power strategy, though he also delved quite deeply into the domestic politics of Sparta and Athens.[3] The most basic point realists make about Thucydides is the historical continuity evidenced by warfare as a means to an end in politics, and that end is either survival or imperial expansion. The real cause of the Peloponnesian War, despite the intricate complexities Thucydides details in his classic history of the event, was the "growth of Athenian power and the fear which this caused in Sparta" (1972: I, 23). Thus would begin realism's emphasis on the structural implications of anarchy and the role played by fear of attack (in a word, the role played by *insecurity*).

Machiavelli's *Prince* reestablished the formal distinction between ethics and politics which appeared in Thucydides and in the Sophists, yet seemed absurd to the Greek philosophical mainstream. In the process, Machiavelli cautioned political leaders within uninstitutionalized authority structures against policies of "benign neglect." The logic of power requires calculated deceit and purposeful application of force (preferably through third parties). Meanwhile, conquest is legitimized for purposes of national unification; again, the survival of the state becomes the dominant end; this is a realist premise in and of itself. Further, Machiavelli was (according to many historians of military science) one of the greatest strategic thinkers of all time; the legitimacy of what would later become known as total war was a matter of faith to him. As Felix Gilbert writes, to Machiavelli, "it was possible to gauge all military measures in relation to one supreme purpose and to have a rational criterion for them . . . this was the beginning of strategic thinking" (Gilbert 1971: 24).

In Hobbes' *Leviathan*, one finds a direct inspiration for realism as a modern worldview. Hobbes' mechanical discussion of human beings as "natural," colliding bodies logically extends to the rapport between "artificial" bodies such as states or other political actors. Of course, Hobbes was not a theorist of IR, concerned with the development of what IR theorists would consider the domestic politics of states. Be that as it may, there are several key Hobbesian themes that would deeply influence realists. Paramount, perhaps, is the fear of death that pushes human beings into a ceaseless consumption of resources. There is also a strong link between Hobbes' assumed amorality of relationships in the state of (pre-Leviathan) nature and the amorality of diplomacy in the international system, though this point is often exaggerated by realism's critics. Other pertinent features of Hobbesian logic include the permanent risk of confrontation between self-interested beings; the assumption of an identifiable national interest; and, surely, the need for a strong state to protect citizens against foreign threats. Hobbes amply discusses how individual sovereignty is to be alienated to the monarch or the Assembly, and how loyalty to the contracted sovereign is to be fiercely maintained (that the Leviathan would deliberately harm its subjects would be an absurdity, since it is the product of popular consent and acts upon its one will).

Acknowledging these roots, one should also recognize that realism's etymology and, arguably, its most lasting influence, derive from historical developments "on the field", i.e. from the *Realpolitik* school in Bismarckian Germany. *Realpolitik* refers merely to a state's prudent assessment of the military balance in an anarchical system,[4] and does not necessarily entail *Machtpolitik*, i.e. power politics in the literal, aggressive sense. However, it is quite clear that the two have been intimately associated in the realist tradition.

Realist politics thus developed a strong following in Europe in the latter stages of the nineteenth century, a time of flowering nationalism. The link between nationalism and the gradual development of a centralizing notion (however nebulous it would prove) of the *national interest* is important in terms of classical realist thought. One might link this development with Rousseau's conception of the general will; Doyle believes Rousseau was the first major philosopher who "identified the *national* interest and made it something more than a slogan" (Doyle 1997: 137). Beyond this democratic origin of the concept, however, a nationalistic one clearly emerged with force in other parts of Europe. Comparing the lives and writings of two key figures in nineteenth-century Europe, Giuseppe Mazzini and Heinrich von Treitschke, one may appreciate how nationalism attracted both centralists and democrats. Mazzini was a key liberal thinker of Italian unification, to be discussed in the next chapter. Treitschke was a fervent German nationalist whose stirring lectures and powerful exposition of far-right arguments attracted a wide audience, won many adherents, and bequeathed a legacy of militarism that lasted through the first half of the twentieth century. Treitschke's *Politics* presented what was then a popular current of thought (authoritarian nationalism) whose historical legacy directly stimulated the re-birth of rational realism in contemporary international relations theory. Realists and liberals differed in their conception of the ideal polity, yet, learning from the British experience, both accepted values of unity, strength, and cohesiveness as a path toward national greatness. In fact, and most paradoxically from our contemporary perspective, nineteenth-century nationalism and the doctrine of *Realpolitik* represented a form of idealism which, today, is associated only with some liberals or, even, socialists. The realism that sounds so "real" today surely idealized power, and while it may have sought inspiration from Machiavelli or Thucydides, it was scarcely divorced from historically specific norms and emotions.[5] Typically, this was a nationalism exulting the success of culture and science, both employed in the grand project of national preservation. The reality of national power, fused with scientific progress, military discipline, mercantilist growth, and bureaucratic autonomy, was to predominate.

With time, this framework for change transformed into a status quo, and realism became the bastion of conservatism for which it is known today. This evolution did not occur abruptly, however. Up until the launching of the behavioral revolution in the social sciences after the Second World War, most pessimistic theorists of international relations cautioned against a rigid reading of reality which might empty political theory of its moral dimension. This indicates an internal dilemma. On the one hand, realists were struggling between their sympathy for scientific analyses of human phenomena and their wariness of mechanistic and fixed assumptions about the nature of politics. E.H. Carr (1946: 5, 148–62), especially, turned away from "pure" realism while attributing moral character to states whose behavior could, presumably, be scientifically investigated. Carr's importance resides particularly in his effort at synthesis between realism and idealism, which, despite its shortcomings, vindicates the power of realism but retains the moral dimension of international politics. While Carr described realism as "the impact of thinking upon wishing," he chastised realists for eschewing the duty of moral judgment (ibid.: 10, 89). His solution was to discuss international politics in terms of "moral states," not as strictly emotional entities, but as actors performing otherwise immoral acts (such as killing) and eliciting a whole gamut of individual emotions which would not otherwise exist (ibid.: 157–62).

On the other hand, however, major military and economic conflagrations were bound to negatively influence the perception of political man/woman. Reinhold Niebuhr (1960: ix, 268) warned that the fundamental morality of the individual could not be projected at the wider societal level. Niebuhr remains perhaps the most lucid exponent of realist thought, and also perhaps the last scholar of reasonable fame to approach international relations from a humanist Christian perspective. His *Moral Man and Immoral Society* painted a fundamentally optimistic picture of the morality of man/woman while upholding the Rousseauran image of society as a state of war – a form of structural analysis which became very popular in later years. Moreover, in both *Moral Man* and in *The Children of Light and the Children of Darkness*, he brilliantly explained how the humane ideals of liberalism will often mask political realities and, therefore, prompt serious errors in policy. Similarly, Hans Morgenthau (1993: 26, 37), the perennial favorite in the classic realist category, asserted the biopsychological roots of power, however checked by existing societal norms or by more general balancing mechanisms at the international level.

By the 1950s, the philosophic roots of realist thought were atrophied with the more general move towards behavioralism in social science. The symptoms were, admittedly, not generalized, especially in Europe where such scholars as Martin Wight, Hedley Bull and Raymond Aron ensured that "international theory" (Wight, 1991) remained political theory. Yet even Wight and Bull were victims of a debatable paradigmatic classification encouraged by positivism.[6] Aron's historical approach was eminently more grounded, which benefited his student, Stanley Hoffmann, perhaps the last true American classicist in the field of international politics.

In the United States, other than in Hoffmann, classical realism was mostly represented in the writings of John Herz (1951) and Arnold Wolfers (1962), neither of whom, however, approached Aron, Morgenthau or Carr's depth and breadth in scholarship.

In the behavioral era, self-described realists largely eschewed questioning about fundamental assumptions regarding man/woman and the state (other than debating the extent of its cohesiveness). This disembodied realist shell, so feared by Carr (1946: 10, 13), turned toward considerations about process, from both systemic and rational-choice perspectives (Kaplan 1957; Waltz 1979; Gilpin 1981). In fact, a seemingly perplexing fusion between realism and liberalism gradually developed. It is true that Morgenthau (1993: 42–3) had vehemently denounced the "peace scientists," intellectuals of liberal leaning who sought to extend the constructive power of science to the social field. Yet the fusion between realism and liberalism seems unmistaken. As social scientists, the peace scientists were arguing that political order could be reestablished through some form of political engineering, and such was the message underlying John Herz' contribution to the field of IR theory. As early as 1951, Herz was calling for an awkward "realist liberalism," where "realistic" means would serve "pure" liberal ideals. Herz' argument is not wholly convincing here, yet he is trying to make a point, i.e. that a society may set for itself lofty goals of growth and progress, but that their attainment requires a realistic assessment of the situation and moderation (or balance) in policy (Herz 1951: 170). The "liberal" component to the argument thus pertains to the process by which policy balance is to be attained – i.e. through the active intervention of the specialist, if not the technocrat.

This realo-liberal mix would find more examples in later works by (especially) Gilpin (1981) and Keohane (1984), all deeply influenced by neo-classical economics and its formal logic. In fact, the evolution of IR theory toward an emphasis on process complicates or, at times, voids the effort at classification. Gilpin and Keohane are two such scholars whose liberalism, for example, is very much entwined with realist axioms: both are fond of an economistic methodology which is typically utilitarian, yet neither would decry an emphasis on power politics in international affairs. Stephen Krasner, a self-avowed realist (Krasner 1974), later associated himself with the regime literature which, in the minds of its proponents, stands outside the realist tradition (Krasner [ed.] 1983). The respected E. Haas quashed the optimism of the functionalist school with a dose of political realism, yet his main career preoccupation remained the study of political networks within international organizations – not a typical realist research agenda (cf. Haas 1964).

In conclusion, and as for any worldview, one should read the defining characteristics of realism as a product of historical evolution and cultural predisposition. From a modern perspective, at the very least, the state-of-war assumption about international politics formally reappeared as realist doctrine in the twentieth century, following several decades of a German nationalist idealism (and in parallel with a liberal tradition, to be discussed later). Realism is usually equated with a necessary, *strategic* deployment of physical force between cohesive, territorialized groups. Yet this worldview cannot be divorced from a practical, cultural experience of force as the expurgation of the soul – a morality in force, but not merely in the defensive (power-balancing, "life-securing") sense. The neo-realist move toward a self-affirmed positivism does not discard the significance of the philosophic roots of realism as a whole, nor does the expansion of the study agenda from Holsti's (1985) marked emphasis on war and its causes to the massive proliferation of topics provided by the liberal-realist regime literature.

ADDRESSING REALIST TENETS

We understand realism as a pessimistic worldview tailored to a world of nation-states in seemingly perpetual conflict. Cooperation is very much part of this global reality, but only as alliance against an enemy, an Other that always exists. Cooperation is instrumental, and not an end in itself, whether or not Hedley Bull's (1977) conception of an "anarchical society" develops in the process. Survival in the system is the key determinant of state behavior. As a worldview (or paradigm, theory, etc.), realism assumes certain perennial features about the world and suggests recommendations to cope with them. In other words, like any perspective on human affairs, it has an ontology, an epistemology, and a program. Is realism then conducive, detrimental, or irrelevant to understanding (and alleviating) environmental problems? And if one can make a statement about the ecological significance of the theory, is this necessarily to say that realism has had an impact on decision-making?

Realist precepts have secured a strong foothold in decision-making circles and in the popular psyche. Defense departments, even the most professional and depoliticized ones, hold tremendous sway within most states, and defense establishments (for obvious reasons) are notorious for embracing a worldview that would guarantee their survival. Military establishments contribute in direct, and indirect, terms to environmental deterioration (see Stoett 1998); and states that may otherwise have cooperated to solve common problems may be inhibited from doing so because of the widespread acceptance of realist doctrine.

There are several key aspects of the realist worldview that may shed light on its ecological role:

1 its ontology of conflict and aggression;
2 its hierarchical component;
3 its emphasis on homogeneity;
4 its materialist worldview;
5 its reductionist epistemology (in the contemporary literature).

We detail these below, yet offer some preliminary comments on each at this juncture. The realist ontology, to begin with, flows from assumptions about human nature and their projection onto larger units – in the contemporary era, states. States are assumed to be inherently death-fearing and power-seeking; these bodies are in perpetual motion and thus are likely to collide. Further, they are rational actors that will take prescient steps to assure that they will survive whatever collisions may await them. This emphasis on fear-from-outside may limit realism's ability to gauge threats from within or from transnational processes, such as environmental decay or longer-term trends such as global warming.

The next key assumption relates directly to the structure of the international system itself. Survival in a conflictual, anarchical world, without any center of power that can serve the role of Leviathan, requires that national "units" be constructed hierarchically. There is a need for a chain of command to ensure the state's survival. We may infer from this precondition that realism is an elitist perspective on human affairs, acceptive of power and authority and the subjugation of some elements of freedom in the name of what is considered a greater freedom, that from external aggression or subjugation. In essence, the security dilemma discussed by Herz and others – whereby efforts to ensure security spiral as states counter-react to each other's threat – justifies a "semi-mobilized" military policy for the country (i.e. not the war culture of Sparta, but still the defense commitment of Athens). Incidentally, if policy is strictly a response to international structure, is this to argue that there need be no direct connection between a realist perspective in international affairs and one's particular ideological affinity in the "domestic" sphere? This is a controversial argument that has been made about Hobbes and could be repeated here (and if it holds, it may be either realism's greatest strength or its greatest weakness, depending on one's interpretation).

Despite this acceptance of hierarchy, which implies difference in social stature and power relations, there is a further strand in realist thought which encourages homogeneity. In fact, sameness may be seen as a recurrent theme in realist theory. History is viewed as a series of recurrences; cycles of power assertion and displacement that result in and from war. While the types of international systems change, from the Greek city-state system to the medieval European state system to the Westphalian system (Gilpin 1981: 41), the essence of global affairs does not. Human motivations are essentially similar, and violence remains a constant source of their expression. Indeed one might suggest that homogeneity is encouraged, as it leads to predictability in behavior. The rules of the game are passed down through the ages. (Here we have some congruence with rational choice theory, though it is usually considered more emblematic of the liberal school.)

All this leads to the characterization of realism as a materialist worldview. Again, survival is paramount, and salvation is achieved by deflecting physical forces through the application of more force. This can only be done with the harnessing of matter, the transformation of nature into means to that end. In this context realism encourages a distinctly utilitarian ecological perspective. Further, in its contemporary expression (yet following early signs in the classical literature), realism supports a positivist, reductionist epistemology which would be rejected by many radical ecologists as part of the problem and not the sole route to solution.

These general themes seem key both in capturing the realist tradition and in providing a basis for an ecological assessment. Indeed, they would appear to seize the crux of realist thought and practice: a worldview suspicious of altruism, content in pursuing a militarized "peace," thus seeking global order through national state control and an effective balance-of-power between major states. Crudely, force is justified according to its utility, and not on ethical grounds. But we should be cautionary about equating realism with aggressive expansionism. Though conquest may be legitimate so as to restore order, the creation and maintenance of empires is often frowned upon as overstretching the capabilities of the state.[7] Further, a realist will not refuse the idea of political friendship, and will indeed accept it so as to improve the ability to reinforce state sovereignty.

Unlike the liberal or critical perspectives, realist discourse will generally not emphasize goals of individual freedom and progress.[8] But it is not inimical to the contemporary language of material growth. If the health of states – their power-base – is today measured in different terms than in the past, then material wealth is certainly a predominant indicator. In fact, peripheral tributes to the power of cultural appeal aside, what emerges from Joseph Nye's response to Paul Kennedy's "imperial overstretch" prognosis is a renewal of realist principles based on material wealth instead of military might (Kennedy 1987; Nye 1990). How such wealth is obtained, and whether or not the process involved is sustainable or contributes to ecological damage elsewhere on the planet, does not seem to grace the radar screen of the realist approach, in particular the structural variant. This would receive immediate criticism from several of the lines of ecological thought discussed in Chapter 2; it is non-holistic, short-term thinking.

Realists insist on the policy paramounce of *security*, but such security is construed in narrow military terms and is susceptible to breakdown from other sources. Critical theorists such as R.B.J. Walker emphasize "the extent to which conventional accounts of security depend on certain assumptions embodied in the principle of state sovereignty" (Walker 1990). This may be gradually changing, as environmental concerns become increasingly obvious sources of insecurity. But it is hardly a paradigmatic shift, as some might suggest, within realist camps. To add fish to the list of things the military establishment is construed to protect from external aggression does not move us far from classic realism's emphasis on territoriality (even if it is sea territory) and the need for a hierarchical institution to maintain it. Geopolitical analysis, which derives principally from a realist tradition, has always stressed the strategic importance of resources, natural and human. It does nothing to suggest we consider their *inherent* value, however. Further, as Deudney and others have argued, adopting the security framework for environmental concerns is one more way in which the latter can be subsumed by a state-centric perspective (see Stoett 1994, for discussion).

Beyond these basic key tenets of realism are several other aspects we feel justify inclusion because they have consequence for ecological thinking. These include the realist ontology of conflict, realism's ultimate dependence on hierarchized political structures at the level of sovereign units, its proclivity towards homogeneity, materialism, and a thesis of immutability, and its reliance on a reductionist epistemology. We discuss each of these in turn below.

The realist ontology of conflict

Realist conflict is deduced from two separate arguments. One strand of realism maintains that the main (survival) impulse of humankind is associational. Political associations, however, are finite and, therefore, exclusionary. Since finite bodies eventually collide, precautions against collision are necessary and justified. This is a structural argument (except perhaps for the initial premise), to which most realists would adhere.

Waltz upholds the argument in its bluntest form, resisting any theorizing on the sources of motivations: "there is a constant possibility of war in a world in which there are two or more states seeking to promote a set of interests and having no agency above them upon which they can rely for protection" (Waltz 1959: 227). Favoring the Waltzian "third image" to its extreme, as Waltz does himself, is of course replete with dangers. Most importantly, it expurgates morality from the study of international politics (the system is a machine), legitimizing deceit and/or the use of force in foreign policy as "a reasoned response to the world about us" (ibid.: 238). It also limits the scope of analysis, unless it is utilized as a presumed context instead of a direct cause of conflict itself.

Structuralist arguments are also present elsewhere, though usually in conjunction with reflexions on human nature. Morgenthau offers a good example, although his logic is, to some extent, contrived. He accepts the basic Aristotelian axiom of association, claiming as well that humans are morally obliged to treat each other unselfishly. Yet the moral obligation seems to struggle with the (selfish) imperative of survival in scarcity: at once, "individual egotisms, all equally legitimate, confront each other" (Morgenthau 1946: 191). Selfishness wins over altruism. But is the ensuing confrontation purely structural, only scarcity-based? Here, Morgenthau introduces another key assumption, namely the desire for power, which emerges (*presumably* within some, but not all, individuals) once survival has been secured, and which ensures the *permanence* of conflict (ibid.: 191–3). Both the target of hegemonic violence and third parties must respond to the "evil" of power: this isn't a mere game of chicken, but the fulfilment of a moral duty – the duty to protect the national interest (Morgenthau 1946: 201–3 and 1993: 12).

This compelling (but not reductionist) influence of structure is also apparent in Herz and in Aron. Herz's insistence on the security dilemma underscores his belief in the "universality" of the struggle for power, which is, however, based purely on mutual fear (Herz 1951: 4) and could be overcome through rational means. Likewise, Aron sees the state system as a state of nature, in which conflict and aggression predominate: "the necessity of national egoism derives logically from . . . the state of nature which rules among states" (Aron 1966: 580); yet this structural logic may not be divorced from the "intoxication of ruling" (ibid.: 73), a condition eventually characterizing state leaders (which, presumably, is as much a product of statism as an outcome rooted in human nature per se).

Carr and Niebuhr, however, do not follow the same line. For Carr, structuralism is not something to be opposed to inherent human drives, for political associations mute the nature in man/woman and instill notions of both power and morality in their evolution (Carr 1946: 95–8). In this sense, Carr is surely the most problematic realist, seeking a "mature thought combin(ing) purpose with observation and analysis," shunning the "exuberance of utopianism" and the "barrenness of realism" (ibid.: 10). Almost necessarily, then, there is no apparent ontological predominance of conflict in politics, international or otherwise. It is not that anarchy induces conflict or that humans seek power. History is the application of power for both moral and immoral ends (and here, realism reaches Carr), but power itself may be displayed in cooperative and conflictual modes – for politics entails both; and international politics is merely an outcome of a large community, a community of states, which might be imperfect and suffer from moral shortcomings, but which is not amoral (ibid.: 162).

Niebuhr, on the other hand, is much more categorical, affirming that "conflict is inevitable" (Niebuhr 1960: xv), that power must be used against power. Yet this is not a reflection of human nature, for people are naturally unselfish (ibid.: xi), but of an immoral society which projects the (thwarted) ego of the human being and, essentially, unites the non-unitable "by momentary impulses and immediate and unreflective purposes" (ibid.: 48). In other words, people can coexist within the large group only when leaders of that group are able to channel the negative energies inherent in this state of affairs into outward, conflictual projection. Universal peace is then clearly impossible. Niebuhr, therefore, and contrary to Carr, does not believe that group relations may be governed by moral rules, and therein lies his realism. But he is not a structuralist, in the mechanical, Newtonian sense, for conflict is reducible to aspects of the human condition, yet necessarily mediated through artificial constructions.

In sum, while almost all key realist authors would agree that the history of humankind is one of constant contention between states, nations, or whatever other groups dominate the political system of the day, this basic conflictual dynamic should not necessarily be construed in mechanical, rigidly structural terms. Another strand of realism, meanwhile, specifically endows political conflict with purpose and cognition: domination is a human need (for, at least, some members of the species); political associations necessarily require dominators; dominators will dominate wherever else domination is deemed possible and/or sustainable (i.e., in the international arena). When Bull, surely inspired by Niebuhr, urges the reader to "recognise the darkness rather than pretend to see the light" (Bull 1977: 320), he expresses what most realists would share: the belief that power drives are innate to and/or systematically developed by, in an institutional context, human beings – at least those men and women who aspire to lead.

The basic point of agreement, then, is that power is omnipresent and relatively visible. Not all realists would necessarily agree with Morgenthau's contention that power is biopsychologically rooted, especially when the assumption extends to all human beings[9] (as mentioned above, liberal and feminist scholars have contended this point with great zeal). But realists will all accept, as an essential premise, that politics is pervasive – that the important things performed in a society or internationally are outcomes of conflictual relationships. There may be the occasional harmony of interests, but even cooperative acts entail conflictual backgrounds. So domination may or may not be natural, but it is quickly actualized in a political system destined to maintain order or build a civilization. Treitschke's Nietzschean exaltation of war, moral and noble,[10] may not be shared as such by moderate realists. Yet the latter will not only treat that political program (adopted by many contemporaries and descendants) as a warning that power and conflict are inescapably tied to the human character, they will also accept that power and conflict can serve a moral function, as we saw above with Carr and Morgenthau.

Second, as stated, the power and conflict uncovered by realists remain visible. One may observe palpable attempts at manipulation or coercion by one identifiable party over another, using classic types of resources (arms, money, status, etc.). While the "power politics" associated with realism tends to suggest an overwhelming preoccupation with military security issues, a true realist will, above all, be interested in this constant background of coercion, irrespective of issues. However, as discussed particularly by Foucault, this approach to power bypasses more subtle (yet even more effective) processes of compliance, and conditions a particular kind of top-down thinking which, ultimately, may be unsustainable. The realist conceptions of power and conflict thus are intricately tied to other aspects of the paradigm, which we will address below.

As a whole, then, realists believe that power drives are natural, that political associations (states or similar finite groups or entities) are natural, and that power drives are served by political associations. In this conception of nature, the strong pursues the weak, the weak is fearful of the strong, and both weak and strong use physical resources to (alternatively) survive or fulfil their natural mission. In fact, survival also animates the strong, who know not only that their life essence is in fighting, but that the weak may grow to be strong and pose a serious challenge. There is an implicit link here between the implications of this ontology and that of the authoritarian ecological thought discussed in the previous chapter: the need for a strong hierarchized unit for survival in a world of opposing Others. This is especially troubling when we consider the historical affinity to seek congruence between nation and state, or between ethnic identification and political autonomy.

This perspective on the ultimate purposes of domination (and, therefore, the preeminent logic of history) distinguishes realists from others. For Marxists, history is the transformation of productive techniques (tools and power structures) for the purpose of elite domination. In capitalism, elitism is class-based and devoted to unceasing accumulation. For most, while elites make history, conflict specifically serves as a means to protect the people. To Marxists, capitalists fight for themselves, and the state fights for capitalists; but for the realist, state leaders fight for the vertically integrated constituency (usually the nation), either to fulfil personal ambitions or to serve the constituency's cause (usually the former).

The realist Hobbesian ontology of conflict makes key assumptions about power needs, fear of death, and the political state of nature. This state of affairs is not to be liked or disliked, but to be accepted and managed – for varying objectives. This ontology is clearly manifest in the writings surveyed here. Still, while some authors labelled as realists will tend to nuance and qualify realist arguments, all within the tradition believe in the fundamental existence of power struggles in a context of anarchy amongst similar types of political units.

Realism and hierarchy

An axiomatic account of history emphasizing conflict and aggression is bound to yield an elitist, hierarchical theory of international politics. The argument is predictable; in fact, for some time now, feminist scholars have maintained that history-as-war/conflict reflects a purely patriarchal reading (e.g. McAllister [ed.] 1982). Many revisionist historians and philosophers, who have researched the role of women in history, may now attest to the systematic historiographical erasure of women's attempts at creating a more peaceful world and challenging the warring culture of patriarchy. The same can be said for many other groups who have been silenced by the dominant cultures of expansionary Europe.

Realism is, of course, fully imbued of hierarchical and elitist conceptions, which extend well beyond the domination of woman by man. Its messianic version, whose normative outlook is not necessarily shared by the twentieth-century political scientist, remains indicative. Treitschke (1963: 11) wrote that "the features of history are virile, unsuited to sentimental or feminine natures . . . the weak and cowardly perish, and perish justly"; not surprisingly, "all democracy is rooted in a contradiction of nature, because it premises a universal equality which is nowhere actually existent" (ibid.: 31). The latter statement, while still wholly arguable, is not at all disputed by modern realists. Aron returns to the broad systemic level (a state of nature, for him), and argues (Aron 1966: 641) that "no international system has ever been, or ever can be, equalitarian." Niebuhr did not praise hierarchies, but was forced to recognize their presumably natural existence (Niebuhr 1960: xiv). Herz, likewise, endorsed the argument, although rather uncritically (Herz 1951:

19); in fact, Herz understood that realism engenders a self-renewing form of elitism, as realist descriptions become prescriptions and favor "the aristocracy" and authoritarianism (ibid.: 29).

So is the "reality" of conflict, then. Nature separates the strong from the weak, as discussed above. But as the strong must remain strong, it must devise a system of accumulation and control which ensures that energies are channelled to a focal point, at the top, so as to protect the vertically integrated entity (the nation-state) against a hostile environment of functional equals. The necessary state, the good state, will not survive without entrenched hierarchies – this is where realist description becomes policy prescription. In its mildest expression, realism merely warns against the omnipresence of power exertion. But the realist logic effortlessly and understandably extends to a theory of omnipresent war and death, which legitimizes the power apparatus.

Realism therefore demands technocracy and centralization: as modern realists, such as Kissinger or Brzezinski, would agree, such are the necessary requirements of "national security." Morgenthau provided a moral defence for "prudence" – a cost/benefit analysis of the requirements for national survival (Morgenthau 1993: 12); yet even his secularization of Niebuhrian principles pales in contrast with the contribution of realist thought to the shaping of the modern military-industrial complex.

In sum, two distinctive points may be stated in a discussion of realist hierarchy. The first is that domination is the *raison d'être* of realist thought. This compulsion of hierarchical thinking is initially rooted in a specific (and biased) understanding of nature – the survival of the fittest; this conception precedes Darwin and is also shared by many liberals and Marxists. Once the analyst (or the activist) accepts a law of nature based on the preeminence of physical strength, then both a conflictual reading of history and the ontological supremacy of violence-organizing forms of association are likewise accepted (or praised, in some cases). This, in turn, condones or vindicates the supremacy of the political association and its elite (knowledgeable, productive, warring) over the individual. In fact, this is performed in two ways: by granting largely unpublicized privileges to the elite (a resource distribution from poor to rich), and by elevating the myth and the glory of the particular abstraction (nation-state, religion, ideology) which already commands legal and moral authority and which can now elicit devotion from the (useful, troublesome) individual.

In its heroic form, realism extols the authoritarian ideal, belittling – but not obliterating – the individual in a quest for civilizing greatness, while still surely marginalizing groups whose genetic make-up positions them at lower levels in the "natural" ranking order. In its analytic and moderate form, as one can see from Aron, realism still asserts the imminence of war and is forced to condone the disciplining power of state authority for purposes of national interest and survival. Here too, then, the individual interest yields to the general interest in what arguably becomes a fictitious symbiosis trivializing human life (and applauding the artificial life of the construct – the national interest, national security, the state itself).

Finally, the second main point to highlight here is that realist hierarchy is also compelled by the vicious circle of description-as-prescription: the world "is" a threatening anarchical system requiring disciplined hierarchies at home, and so such domestic structuring "ought to", "must" be favored. We will not belabor this point, which is a recurrent feature of critical thought in international relations. From a critical perspective, indeed, it is difficult to overemphasize how the current hierarchical order within states was created by men who thought that "survival" and "progress" deserved nothing less.

Realism and homogeneity

Certain ecological worldviews presumably cultivate the flowering of differences in a community, as an essential guarantee for stability and renewal. Realism, however, dictated by its own approach to peace and stability, is forced to uphold the reverse. As history and nature are fundamentally conflictual, the constant threat of war demands a high level of discipline which, as discussed above, hierarchy provides, and which is necessarily accompanied by an ironing out of differences – for obvious purposes of efficiency, predictability, and control.

The argument is not always clearly expressed in realist writings. In fact, one may be misled by some references which appear to fundamentally support heterogeneity. Consider, for instance, Treitschke's rejection of universalism and imperialism (Treitschke 1963: 12). This should not be construed as a form of humanism or anarchism, but as one particular expression of bourgeois nationalism. Treitschke bathed in a glorious epoch of German art and literature, and understood that culture (to which he attached tremendous value) had historically emanated from national strength. Culture had to be respected, and so were national differences to be respected – in fact, both for general cultural reasons and so as to ensure that German glory and honor be regularly purified through war; engulfing Europe or the world would sully German hands and weaken the cultural impulse of humankind.

Seemingly favorable positions on heterogeneity also surface in such authors as Carr and Niebuhr. With Niebuhr, in fact, one may read a type of discourse usually not associated with realism at all. For example, "a genuine universalism must seek to establish harmony without destroying the richness and variety of life" (Niebuhr 1945: 87). This is a qualitatively very different deviation from the realist credo than in the case of the German nationalist. Niebuhr displays his own brand of idealism, which he knows to be unattainable except through a transcendence of realist cynicism and idealist sentimentality, both considered spiritually weak; the divine hand pursues a Christian morality of frugality, justice and mercy, and not national glory or honor. But Niebuhr, the democrat and Christian moralist, still accepts physical force as the necessary accompaniment to political life and immoral society. In this sense, he cannot escape the homogeneity argument: motives of large groups are reduceable to power, and the need to meet power with power compels the predictable, orderly system of production and accumulation which feeds on a homogenizing "rationalization" of society.

Similarly, Carr is rather disorienting for the analyst seeking to uncover his realist face, especially on this particular issue of homogeneity. Carr, in fact, explains the failure of the League of Nations by its own failure to recognize the "diversity" (but also the paucity) of states, whose behavior, therefore, may not be standardized and rationalized according to legal formulas (Carr 1946: 28). Yet one wonders how deeply Carr would commit to an ethic of diversity. He fully accepts a statist ordering of the world which, however infused with moral standards (as Carr seeks to elaborate), supports a power ethic ("aggression is not necessarily immoral"; ibid.: 208) and condones societal efficiency. Thus, for Carr, the policy of autarky would have nothing to do with related objectives of "beauty-in-smallness" and diversity; rather, it is "an instrument of political power . . . primarily a form of preparedness for war" (ibid.: 121).

The search for evidence on diversity within our realist texts does not appear altogether fruitful. One should not expect sweeping statements from these authors, as most of them do not adhere to a pure version of realism. Still, we read Waltz writing of "the illusion that people and cultures are so very much different" (Waltz 1959: 49). Waltz may have a point, yet his writings demonstrate that, for him, an understanding of the world and an acceptable approach to policy both overwhelmingly emphasize a "functional likeness" which likens humanity to a precision machine. Nothing of the kind, on the other hand, would be expected from the realist-liberal Herz. He believes in the "dispersion," the "mitigation" of power (Herz 1951: 176), and also states the need to "fight against the injustices and discrimination inflicted upon minorities and for an improvement of their status" (ibid.: 198). Yet his overall picture is still pleasing to realists (and underscores the modern link between liberalism and realism), for Herz's diffusion of power is merely equated with the separation of powers in a presidential system (ibid.: 176), while his defense of diversity is admittedly only a fall-back position: "ideally," realist liberals ought to pursue minorities' "full integration in the body of the main group" (ibid.: 198).

In sum, realism displays an essential bias against one of the primary goals of some forms of radical ecological thought, the goal of diversity in society (nationally and globally). While textual analysis is a bit erratic here, the homogenizing tendencies of realism may usually be deduced from its "power politics" framework: reducibility of motive (to power quests, physical growth), likening of units (by emphasizing a statist ontology), defense of nationalism, and, often, an aculturalism which reinforces the sense of similarity (no wonder that, of all social science specialists, political realists are surely among the least influenced by anthropological research).

Materialism and immutability

There are at least two more issues of ecological significance characterizing realism. The materialist dimension may be discussed summarily. As ecologists would assert, the ecological principle of finiteness necessarily entails a respect of natural physical proportions for sustainable living. The argument seems well laid-out for an eventual critique of liberal progress, yet applies just as well to realism, whose policies order the constant development of military might. While not all realists would advocate territorial expansion (i.e. imperialism, essentially a self-defeating form of idealism), all will accept it as a political possibility, against which defense is necessary. In principle, a purely defensive military policy might be designed, so as to simply neutralize aggression without retaliation, to convey the message that the energies of attack will simply turn against the initiator; even ecological communities would adopt this scheme, if it could prove feasible. But warfare and war preparation usually blur – and void – the distinction between offense and defense, giving free rein to institutionalized interests to pursue a "status quo" policy of military renewal/growth.

Realist logic condones a materialist framework for the organization of society and the conduct of international relations. Indeed, both classical and contemporary scholarships have constructed a materialist narration of history, centred on a conception of power inevitably defined by (quantitative) measurements of physical capabilities. Such materialism expresses a form of amoral secularism which, however, does not characterize the entire realist tradition. Niebuhr, for instance, is representative of a realist strand whose acceptance of physical power is mixed with an ethic of restraint: in theory, power may be wisely wielded so as to contain violence and sustain moral or religious values. However, there is no evidence that prudent, moral realism can effectively be implemented in a statist, nationalist and/or capitalist context.

The (materialist) emphasis on physical force stems directly from a specific "immutability thesis." Realism heavily stresses all-powerful natural laws which, granted, do leave room for original decision-making (Morgenthau 1946: 220), but nonetheless condition both an ethic of military force (as those laws emphasize conflict) and an epistemological obsession with recurrent patterns. At the most basic level the immutability thesis fixates on human nature, which cannot change and will always be the source of political friction, regardless of societal organization. But even more generally there is a faith in the essential sameness of human history. Gilpin (1981: 7) has summarized the realist immutability thesis by bluntly stating that "the fundamental nature of international relations has not changed over the millennia; . . . (this consists of a) struggle for wealth/power among interdependent actors in a state of anarchy" (ibid.: 211). Thus the importance of the link back to Thucydides, discussed above, and the idea of recurring violent episodes of system change.

In short, we have not – contrary to idealistic expectations representative of the liberal and critical schools of IR theory – transcended the need for violence, or the Hobbesian state of nature, in the IR arena. According to Gilpin, concerns with environmental crises related to the commons will not evoke such transcendence, either:

> Past expressions of neo-Malthusian ideas similar to the current limits-to-growth thesis have not led to the transcendence of narrow circumscribed loyalties . . . national fears concerning over-population and insufficiency of raw materials have led to the most destructive and irrational of human impulses . . . there is little evidence to suggest that mankind has advanced much beyond this level of jungle morality.
> (Gilpin 1981: 224)

Our point here is not to reiterate the substance of the immutability thesis, but merely to underline its presence and emphasis. It remains one of the fundamental purposes of social science to identify alleged constants in the history of humankind; and, while a preoccupation with perennial forces may instill a rather unconvincing form of determinism in theorizing, more and more thinkers are careful to avoid that trap. But one can also argue it suggests a certain fatalism which would be rejected by many environmentalists as extremely limitational. The utilitarian ecophilosophy may have been appropriate during an era of relative natural abundance; what has shifted so many thinkers into alternative modes of thinking has been their perception, right or wrong, of fundamentally changed circumstances, of the realization that an era of unaccustomed scarcity (taken as a literal term) and threats to ecosystemic health is upon us. While structural realism has produced an interesting debate over the implications of shifts in Great Power polarity (see Kegley and Raymond 1994) it has yet to seriously consider the shifts inherent in an even more fundamental context of human affairs – the biosphere itself – perhaps because the immutability thesis pre-empts serious consideration of such developments.

Yet the vast majority of empirical research conducted by realists and liberals alike (or their awkward hybrids, regime analysts) looks precisely at institution-building processes as a consequence of changed circumstances, or the construction of what Seyom Brown has termed "rudimentary international accountability frameworks" (Brown 1995: 257–67). The common realist response has been to emphasize the constraints the international system places on neoliberal institutionalism: problems of compliance, free ridership, state concerns with relative gains/losses (Grieco 1995). The prospect that old cycles based on conflict might be broken, even in a manner as restrained as that envisioned by mainstream neoliberal institutionalism, is viewed as politically naive.

Of course, it may be, but that isn't the point here. Realism offers a tough lesson in the awareness of constraints, even if they are in many cases the consequence of perceptions of the state system or, for that matter, scarcity. But it is also the most limitational of the three schools examined in this book, and that in turn limits its concordance with, and value towards, ecological thinking beyond utilitarian or authoritarian approaches.

Realism as reductionist epistemology

Historically, realism has cloaked itself with a "progressive" veil of objectivity, positioning itself against moralist or religious tradition. Thucydides was eminently modern in this sense, and there is probably little coincidence that he elaborated a systematic approach to war at the very time where another Greek modern, Democritus, sought to popularize a conception of being as a succession of atoms. The parallel with post-Renaissance realism is decidedly striking. Machiavelli had barely tempered the ardor of Renaissance humanism when Descartes and Hobbes were joining hands in formalizing a revolution in political thought, insisting on mechanical cause-effect relationships in a largely despiritized, atomized world.

A general appraisal of contemporary realist texts seems to uphold the reductionist thesis. Yet, again, the evidence is at times contradictory. The "classic" realist authors of the twentieth century can still impress us with their philosophical and historical awareness. Carr, Niebuhr and Aron cannot be lightly accused of the ahistoricism celebrated by Waltz's 1979 volume and, more generally, by the various strands of the "process" literature. Indeed, Richard Falk (1997) has recently placed Carr and Bull – alongside Robert Cox – in what he terms a "critical realist school" which is anathema to the behavioralist orientation. Many such realists, in fact, viscerally attacked what they perceived to be science's misguided appropriation of the "peace problem" – attempts at engineering peace through neat formulas, which would presumably win over a war-torn historical baggage rooted in human nature. Niebuhr (1960: xvii–xviii) jeered such "naive, . . . unqualified rationalists." Carr (1946: 28) insisted on searching for historical contingencies, for the historical power of ideology, in any analysis of political struggle and success.

In *Scientific Man*, Morgenthau declared that "scientism is unable to visualize problems, fields of knowledge, and modes of insight to which science has no access" (Morgenthau 1946: 124). As noted earlier, Morgenthau also distanced himself from Hobbes and Machiavelli by describing their scientism as "merely an accident without consequences" (ibid.: 169). This was the rich and provocative Morgenthau who, while he displayed some sympathy for the social scientific tradition, fundamentally denounced the metaphysical emptiness of scientism.[11]

Still, for all such caution, contemporary realism remains imbued of an almost messianic scientific ethos, rejecting "simple" science but remaining devoutly faithful to a method which, expectedly, seems best suited for the discovery of "what is." The Morgenthau most people remember articulated his foremost principle of political realism – politics is governed by objective laws – and maintained an "autonomy of the political sphere" (Morgenthau 1993: 13) which precisely serves the atomism of Newtonian science. Carr and Aron, for their part, were quite candid in expressing their belief in a science of international politics (Carr 1946: 5; Aron 1966: 6). Bull wrote: "there does in fact exist a close connection between order . . . and the conformity of conduct to scientific laws;" this, conversely, entails "the possibility of finding conformity to scientific law in social conduct that is disorderly" (Bull 1977: 7–8). Even Niebuhr could not shed a rationalist ethic, expressing the Christian confidence in scientific progress: "make the forces of nature the servants of the human spirit . . . the instruments of the moral ideal" (Niebuhr 1960: 256). And Herz's belief in a scientific theory of international politics is easily discerned from his general extolling of science and its promise of progress.[12]

Overall, one might argue that realists have chosen to focus on historical constants and slowly fallen victim to an obsession with recurrent processes. This quest for patterns characterizes most of the recent IR literature, and whether authors label themselves "realists," "neo-realists," "structural realists," or even "liberal institutionalists," they are all concerned with fixed power games (e.g. zero- and variable-sum games, prisoner's dilemma, chicken, stag hunt, and mixed-motive games) whose theoretical underpinning stems from the realist tradition. As Doyle suggests, "Hobbes, with his analytic individualism, would have found the modern analysis of cooperation and conflict found in game theory especially congenial as a way to imagine the possibilities of cooperation and conflict among Leviathans that lack authoritative international order" (Doyle 1997: 118).

Realism, while originally inspired by historical analysis, developed a series of arguments about politics which became easily captured by reductionist social science. Morgenthau urged that historical differences be recognized yet ultimately insisted on historical similarities. Such an agnostic position is indicative of the epistemological pressure within the tradition. This presented somewhat of a paradox for the classical realist who associated modern "expertise" with liberal idealism.[13] However, there is no need to repeat the argument linking realist thought to the shaping of a modern technocratic class, whose problem-solving mind-set favors the status quo.

An ecological assessment of realist tenets

We recall the three main ecological perspectives discussed in Part One: utilitarian, authoritarian and radical. Along this wide spectrum, "ecological practice" will mean different things. The radicals see ecology as linked to freedom – but not the utilitarian freedom to consume. Several authoritarians may accept the necessity of consumption, but will see to its proper management through draconian state measures; alternatively, eco-authoritarianism straddles the radical border through more or less direct flirtation with nihilism. Strict utilitarians will sustain the growth utopia through a mixture of market innovations and conservationist measures, with the help of technocratic intervention. If all ecologists seek to salvage nature from human assault, they disagree on the means and, also, on the threshold beyond which consumption becomes illegitimate. Ecology may be seemingly furthered through anarchy or hierarchy, it may require freedom or ban it, and it may praise high technology or denounce it! One thing is common: sustainability presumably requires "stability," "order," surely "peace" – key words from the IR language and, particularly, realist theory.

As we have suggested throughout this chapter, an eco-radical assessment of realism is bound to be condemnatory. Ecoanarchists would particularly attack what we might term the social-Darwinist basis of realism (i.e. the ontology of conflict) as the initial premise of tyranny: a world of realists will destroy life, for a militarized peace feeds on ecological degradation and cannot guarantee that state leaders will forever remain "rational."

In this light, realist stability and security is illusory. Realist hierarchy, likewise, is to be resisted by ecosocialists and ecoanarchists alike, since their social project is to free the human subject and other life forms through the gradual equalization of status. Realism's conservative orientation suggests that the traditional pillars of "strength" (men of power) should continue to dominate in society, at the expense of women and all other "marginals" (including non-rational beings) who cannot legitimately contribute independently to a disciplined war effort. By extension, this policy of the status quo supports capitalism as the most powerful means of production known in human history. Similarly, realist homogeneity is geared towards an ironing of differences to the benefit of established, powerful classes and cultures.

All in all, radicals will balk at realism as a theory and a policy of control, vindicating the nation-state and the modern military – both of which are indicted for their historic contributions to social repression and reliance on violence as a potential problem-solving device. The radical attack also extends to realist materialism, not necessarily from a religious perspective but surely as a criticism of realism's treatment of "non-useful" matter as expendable in the technological and strategic quest for military superiority. Finally, radicals will question both realism's immutability thesis and scientific reductionism. Realists seemingly foreclose the future (and deny social freedom) by postulating an historical recurrence of violent conflict and studying human behavior through an epistemology of control. Radical appraisals, within and beyond ecological thought, pointedly demonstrate how such philosophical straightjackets are destined to deny the autonomy of subjects and to dissuade alternative thinking. Radical thought is holistic thought and maintains that historical "reality" is repeatedly constructed by the political powerful.

And so the eco-radical argument is two-fold: historical fact is contingent and should not deny a search for emancipatory possibilities, while the "science of IR" is bound to ignore such possibilities as it investigates "natural" patterns through objectification and classification. This is ecological language as much as Ecology stands for freedom – and it presumably can, if the attack on nature can be explained by the control, marginalization and widespread suffering of man/woman in an era of scientific and technological "progress."

That realism and radical ecology would seem incompatible is hardly surprising. After all, critiques of realism already abound within the IR literature, many of which recognize how a militarized order is, seemingly, inherently detrimental to the natural environment; proponents of a "positive peace" particularly signal how a durable world peace hinges on a re-thinking of security, one that targets people and nature through extensive reforms of the political and economic system. As we explain in a later chapter, eco-radicals have clear affinities with Marxists, feminists and even reform-liberals within the critical IR literature. Yet, as not all ecological thought is radical, the question remains: is there such a thing as an eco-realist?

There are evident parallels between realists and non-radical ecologists. One may consider, for example, the overlapping discourse of "realist idealists" and ecofascists: the inevitability of war is here welcome as a means of "purification;" there is a thread linking Nietzsche and Fichte to Treitschke and Hitler. This "ecology of war" may sound downright puzzling, if not abhorrent, yet it does lay claim to some naturalist argument by which the mind and the body are cleansed by the act of war, and by which the nation is allowed breathing space through territorial expansion. However, this is not to suggest that contemporary realist thinkers in the West are pursuing an academic or a political program along ecofascist lines; the coincidence between realism and that branch of authoritarian ecology remains historical.

On the other hand, there is a more perceptible convergence between post-war realist writings and the remaining strands of ecological thought. Realists notably share utilitarian roots with liberals and, thus, are linked to conservationist arguments. Obviously, defense strategists are not likely to be impressed by Brundtland-type attempts at re-thinking security or by any government decision that would cut into the military budget to the benefit of trees or rivers; similarly, the "limits-to-growth" argument would be viewed suspiciously by a military thriving on technological innovation. However, a resource-oriented approach to ecology may still strike a favorable ear within conservative organizations seeking an efficient use of natural resources for security purposes. Some analysts have thus claimed that the military can muster the necessary strength to lead the battle against ecological degradation: hierarchy becomes an ecological wonder, military ingenuity offers the peace dividend. As we noted above, this is a contentious proposition.

As realism is the intellectual defense *par excellence* of the military establishment, it is appropriate to refer to the military as a means to discuss the paradigm's ecological limits and possibilities. From a realist perspective, the military as state actor is not to yield political power to private enterprise. The military-industrial complex is presumably steered by state demands, and so the more congenial link between realism and ecological thought can be found within Green-Leviathan perspectives (which, in effect, are not inimical to conservationism). A Hobbesian ecology would recognize ontological conflict and understand the political dangers of scarcity – whether the scarce resource is oil, food or water. The work of Thomas Homer-Dixon has been extremely popular over the past decade as an assessment of the dangers posed by ecological degradation. The state is here culprit and savior, although Homer-Dixon (1994) discusses various sources of scarcity, beyond the military's own actions. Most of this literature, however, is devoted essentially to maintaining state security and looks not at international conflict but domestic scarcity-related conflict. While it is causally sophisticated, it rarely moves beyond state-centric and, even, authoritarian, response levels to problems. The same can be said for the neoliberal institutionalist literature, to which we return in the next chapter.

To the extent that realism has formed the dominant discourse in IR studies, at one time arguably defining the parameters of the field itself, its focus on great power politics and military conquest and defense has, if anything, curtailed the development of ecological awareness amongst students of world affairs. The closest exception to this came when concern over nuclear testing and weapons fallout forced practitioners of strategic studies to consider the ecological impact of strategy-related behavior. One might argue that the campaign of ecocide unleashed by American forces in Indochina also drew attention to ecosystem health questions. But this was because of growing war fatigue and opposition in the United States, not because realism itself led students to look closely at such topics. Similarly, geopolitics within the *Realpolitik* school has traditionally dealt with the environment as a constraint on strategy.

Ecological thinkers may or may not reject realism's often unyielding premises, but they can still learn much from this most entrenched of perspectives. At the very least this should include an appreciation of the conceptual depth of the ideology of nationalism, and of the central role the state still plays in world affairs today. The security dilemma so often viewed as a comfortable industrial rationale for arms races just doesn't go away, whether it's related to Sparta–Athens or India–Pakistan. Realism's skepticism regarding the role played by international institutions is something else that environmental activists of all stripes should at least keep in mind. Within the more limited field of environmental diplomacy, there is no doubt that an intense knowledge of state interests is essential if we are to understand the evolution of international environmental legislation. Yet it remains difficult to conceive of a realist perspective that is adequately informed of long-term ecological questions.

CONCLUSION

An assessment of realism's ecological credentials will vary according to the preferred ecological perspective. If nature is strictly understood as a resource pool to be properly managed, then realism offers a Hobbesian framework for effective state management of any resource crisis: realists will make the age-old argument that scarcity breeds conflict, and that only strong states attuned to the "national interest" can muster the necessary effort so as to minimize scarcity problems. In other words, state A may presumably deter state B from polluting air or water or from depleting fish stocks. Even "fighting global warming" would seem plausible, although the effort would appear immense and likely unmanageable for such a wide problem.

Overall, however, we may question whether a world of realists would be conceptually equipped to sustain planet Earth in the long run, and thus to further ideals of order, security and peace which are so basic to realist theory – and to IR theory in general. Realists emphasize the immediate threat to security posed by scarcity in anarchy, and argue that rivals must prepare for war so as to avoid war. In the process, one might argue they give carte blanche to the military apparatus and demand unquestioned allegiance from citizens of the nation-state. The war effort requires efficiency in the chain of command and must discourage social pluralism. The result is, arguably, a direct attack on the planet's ecology: not only is the military a voracious consumer of nature, but nature's diversity is not respected at the social level. Realism is a worldview that encourages elite control and that largely ignores the military's own threat to national security in peacetime. This is not to reject the very realistic argument that scarcity breeds conflict, and that Napoleons are just as real as Gandhis, but to argue that in an age of technological complexity and transnational problems realist solutions to insecurity cannot be satisfactory.

Realism is a somber description of inter-group relations in the absence of a formally defined sense of community. History is interpreted as a series of bloody cycles of warfare between collectives, be they city-states or empires or today's nation-states. This naturally emphasizes the "reality" of material power – perpetual physical threats, and attempts to counter-balance them with the development of greater threats. In an analysis of this ontological framework, we have seen in this chapter how realism has certain tendencies toward both utilitarian and authoritarian strains of ecological thought. We now turn to a similar examination of its main competitor for the hearts and minds of IR scholars, liberalism.

Chapter 4

Liberal IR theory and ecology

INTRODUCTION

If it is difficult to tease the ecological implications of realism out of their respective hiding places, liberal perspectives on human affairs appear relatively open to ecological assessments. Liberalism is the historical byproduct of the Enlightenment; as a worldview, it is committed to the rational harnessing of nature in the quest for human freedom. Though human freedom is associated with citizen relations vis-a-vis the state, it is also closely linked with economic growth, the ability to prosper and escape the confines of poverty and reliance on nature. Nowhere else is the link between IR theory and utilitarianism as clear and forceful. Further, the liberal ethic would reject the authoritarian ecological thought introduced in Chapter 2 on the grounds of the latter's anti-individualistic orientation. Radical ecological thought would also receive scant praise from most liberal thinkers, though there may be interesting exceptions here related to a libertarian ideology and ecoanarchism on the one hand, and variants of welfare liberalism and ecosocialism on the other.

Liberalism's explicit stress on individual development suggests an atomistic ontology, but one different from realism's state-centricity. For liberals, "the only real measure of progress is in fact freedom, and by freedom is meant the development of the responsible and autonomous self" (Manning 1976: 25). But, as an emancipatory framework, liberalism is naturally sensitive to the critique of radical ecologists: whose "freedom" is pursued by liberal thinkers, exactly? How does individual autonomy provide environmental security? As a program dedicated to progress, liberalism is also aligned with the forces of science, technology and capital, spreading cultural norms and shaping the gradual evolution of a global civil society. Despite its accent on individual autonomy, then, the homogeneity resultant from the widespread adoption of liberal philosophy, or more generally what is often termed Westernization, promotes sameness. In short, what Doyle and others term "Commercial liberalism" both authoritarian and radical ecologists might consider anathema to the principle of differentiated sustainability.

This chapter will proceed much as the previous one, offering a brief discussion of the philosophic roots of liberalism as a branch of IR theory, then moving this ontological premise into a more nuanced discussion of environmental considerations. This exercise is vital given contemporary "problem-solving" approaches. If we place regime analysis within the liberal camp – or, for that matter, accept it as an awkward or alternately powerful synthesis of realism and liberalism – then it is nothing less than axiomatic in any attempt to bridge the gap between the two disciplines of ecophilosophy and international relations theory. Most of the empirical work that has emerged in recent years dealing explicitly with environmental issues and international relations deals precisely with regime formation (see Haas *et al.* 1993). It is liberal institutionalist in character, and though non-state actors play important roles, they are for the most part secondary players. This stress on interstate interaction should not force us to neglect liberalism's central focus on the individual, however.[1]

Beyond the regime literature, one may identify a much stronger current, which has swept the corporate and academic world with great speed: this is the age, we are often told, of *globalization*. The ideology of globalization is decidedly liberal in character. It is viewed as progressive in that it improves the livelihood of countless individuals; it is seen as logical given the irrationality of trade constraints and excessive governmental interference in the marketplace; and it moves us toward a universal value-system based essentially on Western, middle-class values. All of which makes realists and many critical theorists alike cringe, the former with skepticism, the latter with outright hostility towards this latest modern project. This isn't the place to assess the empirical validity of claims to globalization's ultimate impact – or even to the claim that there is anything distinctly new about it. But as the most recent manifestation of liberal IR thinking, it deserves treatment in the context of a discussion on ecological thought and world affairs. Though the prospect of increased cooperation on environmental issues through institutional design and growth may please many environmentalists, it may be coming as part of a package deal that in fact decreases heterogeneity, increases extractive activity, and emphasizes technocratic problem-solving to what are in essence political and, even, philosophical dilemmas.

THE EVOLUTION OF LIBERAL IR THEORY

There is a tendency to describe the rise of liberalism in IR theory discourse as a *reaction* to the sustained predominance of realism. This approach is taken, for example, by Ole Holsti, who focuses a recent discussion of liberal theories on two common denominators, both of which are challenges to the core principles of realism. Liberalism distinguishes itself "by asserting that inordinate attention to the war/peace issue and the nation-state renders [realism] an increasingly anachronistic model of global relations" (Holsti 1995: 43). The implication here, and it is a popular one in IR texts, is that realism was fine in the past but it simply hasn't kept up with changes in the real world (it is no longer realistic!); liberalism has been invented by recent empirical necessity.

Such a stance is problematic for several reasons. It minimizes the substantial and lasting impact of realism's philosophic tradition, discussed in the last chapter. Further, the challenges posited against realism in the 1970s by the complex interdependence school (Keohane and Nye 1977; Rosenau 1980; Mansbach and Vasquez 1981; Scott 1982) did not necessarily reduce realism to a tired anachronism. One might even argue they simply reinforced realism's state-centric perspective while broadening the issue areas up for discussion. But more to our present point, liberalism has a profoundly influential cluster of philosophic roots that are ignored if we begin where complex interdependence does. Liberalism may have gained acceptance in mainstream IR theory and research as a consequence of growing frustration with the realist paradigm, but it was present, and influential, long before realism became the standard approach taken by political scientists.

The liberal worldview assumes that universal peace is possible, if only human beings could explore the reasoning capacities that they all share so as to devise effective mechanisms of international governance. These reasoning capacities are more or less equally held by all humans, though many of us are held back by lack of education. State leaders are no exception. Therefore, international law may be constructed so as to regulate international exchanges and allow for the efficient production of goods to be enjoyed by the majority. Human beings are not saints and they are surely self-interested, but they remain essentially good, seeking peace and stability without recourse to arms. This differs from realist constructions of human nature, though Hobbes and others did believe that monarchs were capable of rationality and would avoid the costs of war if possible.

Though it is often used as a realist model for interstate distrust, we might employ Rousseau's stag hunt analogy here for a liberal moment as well. As the hunters enter the wood, they face a dilemma. Should they take the chance that their fellow hunters will display solidarity and finish the hunt until a stag is taken, or should they grab their own hare and head for home? Even though they would benefit more from sharing the stag, the thought of a hare as guaranteed reward for the hunt is obviously a tempting one. Liberals would tend to believe that all the hunters are fully capable of realizing the greater benefit from the stag, and therefore from cooperation. Realists would tend towards the opposite interpretation, given what they consider the anarchic context in which such decisions must be made. Or, as Michael Doyle puts it:

> Accurate information, transparency, is crucial to cooperation: If all the hunters truly recognize that all the other hunters recognize the superiority of shared venison, cooperation should be forthcoming. The 'dilemma' arises because under anarchy, hunters wonder whether these understandings can be made clear to the dim-witted or enforceable against the perverse.
>
> (Doyle 1997: 121)

Liberals would tend to believe that dim-wittedness and perversity are both functions of inadequate institutions; that the fuller development of the individual (via the intellectual fruit of the Enlightenment) can replace them with rationality and self-enlightened conformity.

This leads ultimately to the presumed possibility of peace attained through rational thought and creative institution-building, as opposed to counter-balances of force. Peace need not be, as realists would insist, a mere break from war. It is an attainable end in itself. This is not to say that liberals are pacifists, for they do recognize that, in any given epoch, self-interest will unleash passionate quests for power and wealth on the part of some individuals, thus threatening global stability. Therefore, liberals accept the necessity to prepare for war and to fight so-called just wars against the pariahs of global society – using competent authorities, aiming at swift victory, and adhering to strict norms of conduct during war (e.g. sparing civilians and prisoners of war). The declared political purpose of liberalism is the rehabilitation of the common individual, rescued from the clutches of the church and aristocracy; liberals recognize basic human rights, and uphold those rights through a discourse and policies defending reason (science, education) and material growth (utilitarianism, free trade, technology). A globalized peace is necessary to the freedom of the individual; but perhaps this is a hegemonic peace dictated by liberal culture (individualist, pluralist, materialist), with questionable ecological credentials.

A review of liberal IR theory may begin with a reference to Immanuel Kant's discussion of perpetual peace (Kant 1983 [1795]), surely the better known amongst the various projects for perpetual peace expounded by liberal scholars in the eighteenth and nineteenth centuries. Kant has become known as the great cosmopolitan liberal; he specified certain conditions and proposals which, actually, many contemporary liberal authors do not necessarily follow strictly. He believed that peace was ultimately a function of the inviolable state: states would be secure only in the knowledge that they were considered inviolable by other states. Collective security, rejected by realists as an adequate form of protection, is much more a liberal derivative than a Hobbesian one. Kant also held that republicanism was an essential condition of peace. This refers to a political system where there is a clear separation of powers, and where elected representatives live under the rules they themselves pass. Of course, there is a wealth of recent empirical work that seeks to determine whether or not liberal democracies are more or less likely to go to war with either each other or non-liberal states; we return to this controversy later in this book but mention it here to stress the continuity of Kantian thought along these lines. The implications of a positive co-relation between liberal democracy and peace are inspiring for some, but potentially imperialistic for others.

Contrary to some interpretations, perhaps based more on transnationalist zeal than an accurate reading, Kant did not envision his eventual peaceful world as one united under a single political jurisdiction. On the contrary, he "imputed autonomy to the state and inferred from this a duty of nonintervention" (Onuf and Johnson 1995: 191). At the most, we might interpret Kant's *pacific federation* as just that, an arrangement whereby political integration has made intra-systemic warfare obsolete, but hardly a global Leviathan with far-reaching power. More importantly in the modern context, Kant linked peace and commerce; the freedom of the latter made the former possible. In a famous line Kant asserts: "The *spirit of commerce* sooner or later takes hold of every people, and it cannot exist side by side with war." Of course this statement does not sit well with the critical theorists discussed in our next chapter, many of whom would argue that commerce can in fact be a primary cause of imperialistic war and not its nemesis. Yet the basic Kantian message did reach and has been imprinted on succeeding generations of IR scholarship: war is an evil which, if sometimes necessary, may be largely prevented through creative engineering – both in a structural sense and at the level of public opinion (through rational education).

While Kant offers the necessary philosophical background, two nineteenth-century figures, Richard Cobden and Giuseppe Mazzini, seem particularly important in the elaboration of a liberal approach to international affairs. Both made their mark as writers and politicians. Cobden is important for the lucidity of his writings on free trade and his impact in repealing protectionist legislation in England (especially the Corn Laws), thus setting the stage for the golden age of British imperialism. The ideology of free trade, so basic to liberal thought, is inextricably and historically related to the academic and political work of Cobden. Likewise, but a few decades later, Mazzini exerted a tremendous dual influence on the spread of liberal ideas. He is better known for the political movement of Italian unification which he founded and piloted, at times in exile. Yet the movement was squarely constructed upon "modified" liberal principles, passionately defended by Mazzini in his many writings. Mazzini's essential contribution was to popularize the Rousseauian idea of organic nationhood as the vehicle for human freedom and social peace. At the same time, this was to form the basis of twentieth-century state-based internationalism: a modern celebration of the Grotian ideal, fully endorsed by contemporary liberal theorists of international relations as a background for process-oriented arguments derived from social-choice theory.

The early twentieth century belonged above all to US President Woodrow Wilson, whose universalist outlook and concurrent political power helped construct the League of Nations. The first three decades of the century were quite propitious to liberal ideas, especially in the wake of the First World War, and while no single book by Wilson may summarize his thoughts on IR, he did leave numerous speeches which may be used as evidence (though this chapter will exclusively rely on the Fourteen Points Address). The validity of national self-determination in the postcolonial era within a society of autonomous states is still a central liberal premise, though it is often tested by the complications of actual separatist groups. On a strictly academic level, however, the liberal argument of the early century was expressed most reputably in 1911 by England's Norman Angell. His *Great Illusion* is a *bona fide* classic of IR theory, provocatively and cogently arguing the case of interdependence in modern global society; one might even argue the contemporary interdependence literature is no more than a refinement (if not merely a restatement) of quite an old text.

The functionalist work of David Mitrany followed, some thirty years later. The belief in a science of peace, so decried by Morgenthau, is perhaps most celebrated here, at least in this century. Mitrany's importance is not gauged by the list of disciples to his work or the acceptance of his functionalist approach; in fact, the technocratic, depoliticized argument of functionalism was embraced only in part by the founders of European integration, who did not subscribe to the optimistic automaticity of functional cooperation. However, the functional logic, as defended by Mitrany, most certainly played a key role in ushering in the new era of international organization, turning to technical experts and to codified law for a solution in containing conflict (see Groom and Taylor 1975). Functionalists envisioned such integration as a process arising out of technical cooperation amongst nation-states, and in the behavioralist era (1960–1970s) neo-functionalists stressed the role of mutual self-interest in the construction of institutions whose success would "spill over" into other areas of interaction. In the development of enlarged political community, then, form should follow function: international organizations should be constructed according to the specific needs they could satisfy for the citizens of states, and eventually those citizens would come to realize that their loyalty to the nation-state was itself misplaced. The European Union has been the traditional source of empirical inspiration for functionalism and neo-functionalism; regional economic arrangements are heralded as embryonic political communities, since a "regional market's institutional machinery, its harmonization of economic policies, and the spillover effect of its successes may help create an awareness within the region of the advantages of the integrative process" (Riggs and Plano 1988: 290).

Functionalism evolved through neo-functionalist, integrationist and institutionalist variants. Neo-functionalists argue that, in some cases, self-interest will be best pursued by such cooperation, which will then spill over into other areas, perhaps more divisive initially.[2] The key author in the neofunctionalist school was Ernst Haas (1964) to whom the ("realist") political qualification of functionalism is credited, and from whom work on "epistemic communities" may be said to derive. Karl Deutsch focused his work empirically on the communicative dimension of integration and international community-building. He contributed to the larger school of integration theory, while Robert Keohane's thirty years of scholarship have rekindled the notion of interdependence in international relation theory and restored the importance of institutional analysis. The more widely read literature detailing the processes of global eco-politics today derives from a combination of these sources, which – though it is clearly too broad to label them all liberal – have a common focus on rationality and institution-building.

To this list must be added one book, one name, and one project. The book is the plan proposed by two legal specialists to reform the United Nations. Known as the "Clark-Sohn Plan" (1958), it is often quoted as an example of the potential of peace-through-law. Some aspects of the plan are, in fact, progressive enough to warrant its classification into the next chapter of this book; indeed, the plan was revolutionary enough to be dismissed out of hand when presented to the Eisenhower administration, and may be said to have ecological significance (from a radical perspective) in its commitment to extensive demilitarization and economic equalization. Nonetheless, it is introduced here in virtue of its continued support for powerful global institutions.

The name is that of James Rosenau. Admittedly, Rosenau's eclecticism immensely complicates any classification exercise, and his name ought to reappear later in this book as an example of the "new" scholarship in international relations. However, the liberal bent in Rosenau's writings is unmistakable; a close reading of perhaps his most important book, *Turbulence in World Politics* (1990), reveals a tendency toward a positive evaluation of the cosmopolitan impulse.

The "project," finally, refers particularly to the World Order Models Project (WOMP), but may also extend to the general tradition of peace studies and peace research. The contemporary search for a "peace formula," so criticized by Morgenthau, dates back to

Lewis Richardson's mathematical work in the 1920s (Richardson 1960), and has since influenced peace research in pursuing a scientific understanding of the conditions of war and peace; its "liberal"-positivist character, quite transparent, *inter alia*, in Kenneth Boulding's oft-quoted *Stable Peace* (1978), has trickled down to the regime literature of the past twenty years.

The WOMP, however, is very different. Launched in 1968 by an international community of scholars, its purpose was (and still is) to understand peace in the broadest, positive sense, and to devise blueprints (and, sometimes, transition scenarios) for a better world, where objectives of equality, non-violence, justice, and ecological soundness, may all be realized in the global system. However, a liberal inclination is clearly visible in the writings of key authors linked to the Project; these are the WOMP caveats which will require discussion in this chapter.

ADDRESSING KEY LIBERAL TENETS

The main lines of liberal thought should already be manifest from the above discussion. Can ecologists identify an ontology, an epistemology and a political programme which they may want to question? What is the path through which liberals (in IR and elsewhere) secure objectives of peace, stability and progress? In what sense is the liberal order congenial to (or incompatible with) ecological precepts?

We may pursue three lines of inquiry, all of which are related. The first is the universalizing dimension of liberal thought, surely the key defining characteristic for purposes of theorizing in IR. While most liberals would rather insist that their worldview is squarely pluralist, one could also read in liberalism a project (not always deliberate) of international cultural convergence (since all human beings are equal as reasoning beings). The second is the utilitarian basis of order: peace and stability hinge on material growth, dictated by the power of private enterprise and fuelled (not hindered) by global interdependence. Third, and perhaps paradoxically, is the active role of the state in enlightening society; beyond the rule of law is the rule of the technocrat (and of the educator), urged by functionalists concerned with the politicization of technical issues in a context of mutuality of interests.

In sum, aside from (and perhaps even embracing) the economistic streak in their argument, many liberals will display necessary characteristics of an "idealism" traditionally opposed to realism. Differences may be ironed out through reciprocated discussion; contact spreads understanding; human beings are intrinsically good (society does not change that); politics is about effective management, not power quests; good laws will be effective in maintaining order as people are educated in (or socialized into) accepting them.

Universalism in liberal IR theory: roots

For some ecological thinkers, the principle of diversity is a cornerstone. What is the liberal position on this issue? The historical purpose of liberal thought was to recognize the inviolability of the individual, and to uphold a productive and progressive social system set in motion by the (more or less regulated) competitive energies of pluralized forces. The pluralistic streak in liberalism does contend, however, with the universalism embraced by the tradition. Part of the liberal project, of course, is to build bridges between nations and cultures, so as to realize both the utopian conquest of nature and material security (through global comparative advantage) *and* the more romantic ideal of human unity. The paradoxical liberal reflex is to be wary of differences "in the big scheme of things" (i.e. to posit an ontological harmony of interests), while encouraging differences within that context of harmony (for purposes of efficiency and renewal). Too much differentiation is either paralyzing or conducive to multi-edged conflicts; too little is, simply, stultifying.

In the next two sections, however, we stress what McKinlay and Little (1986: 44–5) have described as the inherently globalizing character of liberal thought. The goal of unity elicits both geographical convergence and presumably natural human impulses toward rapprochement. The liberal ontology is thus one of cooperation, with definite implications for cultural and political unity at the regional and planetary levels. Liberal unity is usually understood as a convergence of atomized individuals, yet the theme of unity has also served as a springboard for a modified version of liberalism, focused on the nation-state. As we will see below, the nineteenth-century romantics, personified here by Mazzini, closely linked the fate of man/woman to that of the nation, paving the way for a cult of state sovereignty which, today, is the (unemotional) flagship of the "institutionalist" literature.

Rather predictably, the universalist theme runs consistently through the various writings surveyed; it is either advocated as a norm or "read" as part of the unfolding "reality" of world politics. As mentioned, the value of global unity is quite reflective of a "moderate," optimistic approach to power. Cobden (1903: 206), for instance, while not directly claiming a cooperative nature for human beings, condemns the traditional advocacy of balance of power as overlooking the possibility of peaceful growth. Cobden's approach to power and to human nature is that of an idealist – yet one whose idealism is much more a function of liberal rationalism than of Christian ethics. Thus, we may read, on the one hand, that:

> This "rule" [the balance of power] would, if acted upon universally, plunge us into a war of annihilation with that instinct of progression which is the distinguishing nature of intellectual man. It would forbid all increase in knowledge, which [. . .] is power. It would interdict the growth of morality and freedom, which are power.
>
> (Cobden 1903: 205)

Yet the rationalist assumption appears more clearly further on:

> [The need for defense] arises from a narrow and imperfect knowledge of human nature, in supposing that another people shall be found sufficiently void of perception and reflection – in short, sufficiently mad – to assail a stronger and richer empire, merely because the retributive injury [. . .] would be delayed a few months by the necessary preparation of the instruments of chastisement.
>
> (Cobden 1903: 235)

"Rational peace" is thus the cornerstone of the Cobdenite approach to international relations, whereby the possibilities for global unity are found in the individual. Part and parcel of an ideology of growth (for which no apologies are offered),[3] yet not quite the cultural argument proposed by later thinkers, the Cobdenite scheme reaches for "the best" within humanness, unfettered by the reactionary demands of governments – artificial entities, if anything. Cobden (1903: 216) thus enunciates his maxim: "As little intercourse as possible betwixt the *Governments*, as much connection as possible between the *nations* of the world." In later times, liberals would not be so strict as to reject intergovernmentalism as a legitimate path to peace, growth, and security. But in the early nineteenth century, intergovernmental contact was usually associated with war.

Of all the authors consulted here, Mazzini was the only one prepared to embrace war with a passion. While Cobden might have accepted a just cause, his utilitarian ethos would have him write that "(o)ur object has . . . been to deprecate war as the greatest evil that can befall a people" (Cobden 1903: 194). As discussed in the previous chapter, Mazzini (1945: 92) would appear to echo the most frightening calls of Germanic heroism: "War, like death *is* sacred; but only when, like death, it opens the gates to a holier life, to a higher ideal. I hail the glorious emancipating battles of Humanity." Yet the key term is precisely that of emancipation, of revolution directed by the holy nation: Mazzini's totalizing liberalism seeks a new world order, to be forged from below against established castes. Universalism easily finds its reserved niche. Witness first the economic realist: "(H)umanity is one sole body. Think you that it will suffice to improve the government and social conditions of your own country? No, it will not suffice. No nation lives exclusively on its own produce at the present day" (ibid.: 117). Then the moral philosopher:

> I abhor that which is generally called politics . . . I abhor everything which separates, dismembers, and divides; everything which establishes different types independently of the great ideal to be followed; everything which implicitly denies human solidarity . . . (T)here is only one real scope: the *moral* progress of man and humanity . . . Italy matters little to me, if she is not to accomplish great and noble things for the good of all.
> (Mazzini 1945: 117–18)

Ultimately, with the simultaneous passing of the nineteenth century and of Mazzini himself, a particular brand of universalist theory would effectively come to an end. Many aspects of Mazzini's liberalism would not be echoed by mainstream theorists, in his century and ours, for he expressly rejected the utilitarian perversion of liberal thought[4] while grounding his critique of realism on specifically moral, religious grounds.[5] Yet Mazzini did contribute substantially to the liberal current of intellectual history. As we will see below, he remained a product of the Enlightenment, fully endorsing a progressist path to peace based on popular education. International unity, however, formed the crux of his belief. He may not have suggested that such advocacy be used to translate imperialistic, homogenizing designs. However, by legitimizing the nation-state as an instrument of the good, he did open the door to influential arguments in the liberal IR literature, globalist and statist in kind.

Mazzini's legacy was much more apparent (even if still in a partial way) in the writings of Wilson than in those of Angell. Beginning with the latter, however, we see a resuscitation of the Cobdenite argument and the renewed exposition of classic liberal views on human nature and relations. The influential concept of interdependence is clearly laid out in *The Great Illusion* (thus decades before the "complex interdependence" school of the 1970s), derived specifically from economic observations and reflective, presumably, of the self-interested nature of humans (Angell 1911: 52–77). Thus, while Angell does not portray human beings as altruistic or necessarily good, they do appear as reasonable creatures who should understand the benefits derived from cooperation – mutual progress and peace. Society is, then, both necessary and constructive, at all levels of aggregation: herein lies the foundation of universalism.

Angell's views on cooperation are particularly interesting, as he articulated, in powerful language, undoubtedly one of the first "psycho-historical" arguments for peace in the international relations literature. Consider this particular rejection of realism:

> We are all . . . losing the psychological impulse to war . . . How, indeed, could it be otherwise? How can modern life, with its overpowering proportion of industrial activities and its infinitesimal proportion of military, keep alive the instincts associated with war as against those developed by peace?
> (Angell 1911: 205)

The eruption of the Great War, shortly after the release of the book, would seem to indicate the fallacy of Angell's universalism,[6] to which he gave status of natural law (ibid.: 246) and which he built on questionable assumptions about morality and power. Indeed, the point is not to criticize unduly Angell's optimism about peace; he refused to believe that political divisions and ensuing violent conflicts are the natural fate of humankind. The problem lies in Angell's facile alternative:

> The greater economic interdependence which improved means of communications have provoked must carry with it a greater moral interdependence, and a tendency which has broken down profound national divisions [. . .] will certainly break down on the psychological side divisions which are obviously more artificial.
>
> (Angell 1911: 314)

Unity amongst *a priori* similar beings is then easily stimulated by apparently neutral technological forces. This would be echoed by Rosenau eighty years later: "television overall is politically neutral, merely a channel through which the cascades of postinternational politics pulsate" (Rosenau 1990: 346). The process is not explicitly teleological, yet there is an unmistakable impression of a world evolving "naturally" towards unity – a welcome unity of individuals (not of states) sharing the bounties of nature and frolicking in the advances of science. Angell would never have advocated an enforceable unity under a world state, and was careful to address the problematic nature of military power (Angell 1911: 268). Yet he surely underestimated what powerful armies and navies can do to ensure a cultural hegemonic victory, posing as a moral good for humankind (ibid.: 69).

Angell's universalizing vision of peace was meant as the logical extension of a process of interdependence based on non-military means – namely, financial credit. Angell believed in an economics of peace and understood the rising influence of non-state, transnational actors (namely, financial institutions). However, while he did not seek formal, centralized political structures, he offered no suggestions on the means to safeguard human diversity; he may have advocated decolonization, but did so purely according to a cost-benefit analysis (favoring Britain, obviously, although he would have argued that the financial security of Britain was India's gain – and the world's) (Angell 1911: 35).

Of course, none of the above comes as a surprise; this is a mere expression of the classic universalism expressed by the British school of IR. Could the Cobdenite view sustain a legacy in the twentieth century? Should we have expected IR theorists to continue defending the "fundamental reality" (and goodness) of a benign, largely stateless, and convergent world of traders and bankers? Our discussion of realism does provide a good part of the answer. As for liberalism, it did not die, of course, but its globalizing outlook wavered between statist and non-statist poles. This could be construed as a dichotomy between a confederal advocacy of international organizations and a functional/cultural path to a world government-society. Yet it should not conceal the solid endorsement, by all liberals, of international law and freer trade as paths to peace.

Woodrow Wilson's famous Fourteen Points Address to the US Congress (January 8, 1918) reflected a much toned-down – "realist" – version of Mazzini's world-order vision based on the nation-state. The birth of the League of Nations may well be traced to Point 14: "A general association of nations must be formed under specific covenants for the purpose of affording mutual guarantees of political independence and territorial integrity to great and small states alike" (Link 1984: 538). And so the commitment to the sanctity and equality of nations may be inferred from the other Points, most of which called for the evacuation of occupied territories and the creation or strengthening of nation-states, such as in Poland and the Balkans (cf. Points 6–13; ibid.: 537–8). However, as the core of Wilsonian peace is located in free-trade policies, it is doubtful that a formal recognition of state equality could provide the basis of a diverse world. In fact, one should not forget Wilson's commitment "to fight and to continue to fight" (ibid.: 538), so as to impose the *definitive*[7] American version of "unity-in-diversity."

The "Grotian" view embodied in Wilson's Points does converge, to some extent, with the functionalist school; as we will elaborate later, both approaches are firmly based on the principle of peace-through-law and on the positive role of international institutions. But Grotians and functionalists do not necessarily convey the same globalizing message. Admittedly, functionalism is not a monolithic school. It evolved from the writings of the "Red Professors" (especially Harold Laski) to those of Mitrany, from an essential concern with capitalist exploitation – which sharply contrasts with Wilsonian liberalism – to a narrower focus on inter-group politics.[8] Yet functionalists, from Laski to Mitrany, appear united in proposing a social internationalism qualitatively different from the Grotian model of (formal) state equality. Does this entail a world state? The functionalists, seeking international peace through the efficient delivery of "services," would only rule it out as an unpractical alternative (Laski 1967 [1925]: 230).

The important point, however, is that the state system gets in the way, politicizing exchanges to an unbearable extent. Achieving the ultimate goals of freedom and happiness, will depend on which social services Mitrany defends as "practical tasks" (Mitrany 1966: 33). Surreptitiously, or perhaps unwittingly, the idealist Mitrany slips into a politico-cultural form of imperialism, wishing for the day where small states will surrender some of their formal equality for the haven of efficient services: "All the efforts to devise an international system, all the demands for restraining national sovereignty, center upon this issue of how to bring about the voluntary and progressive evolution of world society" (ibid.: 35). Frontiers are to vanish, the functional approach "overlaying them with a natural growth of common activities and common administrative agencies" (ibid.: 62–3). The commonalities in question are not detailed specifically, yet, if the logic is applied globally, they are sure to convey Western modes and values.

Mitrany was committed to finding a formula which would secure unity in diversity (ibid.: 27). The goal is indeed essential. Yet what Mitrany did achieve is to demonstrate, once more, the inherently totalizing objective of liberal thought. Liberal IR theory, more or less subtly, would continue pursuing the task after the Second World War, inspired by both the Grotian realism of Wilson and the socio-technical approach of Mitrany, yet also heavily influenced by the rise of behavioral science.

Liberal universalism: the contemporary literature

Behaviorism did indeed tinker with the liberal school. The universalizing arguments to be gleaned from neo-functionalists, integrationists, and regime theorists do not appear as straightforward as in the writings of idealists, for those particular strands of scholarship abandoned much of the prescriptive intent of IR theory. The postwar "liberal institutionalists" (to use a broad generic term) have been identified as liberals largely by default, interested as they were in the so-called "low politics" of economic and social relations. Thus Keohane (1989:10) admits candidly that "although I subscribe to [the] belief [in individual freedom], this commitment of mine is not particularly relevant to my analysis of international relations." Liberalism is thus squarely associated with institutional process, with the *mechanisms* by which power bows to the forces of law and by which institutions shape political behavior: "liberalism [. . .] serves as a set of guiding principles for contemporary social science; [. . .] it stresses the role of human-created institutions in affecting how aggregates of individuals make collective decisions" (ibid.), or, in Mancur Olsen's terms, in affecting the search for "politically feasible ways to increase the incentives for collectively rational behavior" (Olsen 1971: 874).

Returning to the question, then, how does the positive (non-normative) liberal theory of IR contribute to universalist arguments? Overcoming the stag hunt dilemma remains central. As process-oriented literature, very little is advocated specifically; one must usually recognize an indirect impact, as legitimizing the use of particular axioms, concepts, or methods. Consider, for instance, the underlying interest in the issue of peace, understood as the regulation of interstate conflict in some confederal context. Some declarations by key authors are worth noting. Deutsch *et al.* (1957: 3), for instance, stressed their normative concerns at the outset of their landmark publication: "We undertook this inquiry as a contribution to the study of possible ways in which men some day might abolish war; . . . we are seeking new light with which to look at the conditions and processes of long-range or permanent peace." In the same vein, Etzioni (1965: x–xi) opens his oft-quoted book as follows: "the rise of regional communities may provide a stepping-stone on the way from a world of a hundred-odd states to a world of a stable and just peace. Such an achievement seems to require the establishment of a world political community . . . [yet] not a world empire."

Haas, for his part, is much more reserved – much impressed, in fact, by the daunting obstacles set by political conflict. There is no grand peace formula for Haas, not even a longing for peace; at most we can welcome some level of integration through painstaking engineering. And the same realism, the same caveat stressing the culturo-economic background of homogeneity for integration, may be found in the "transnationalist" literature pioneered by Rosenau (1969, 1981) and Keohane and Nye (1971) which laid the basis for the regime literature. In fact, one can argue that the question of peace slipped slightly on the list of liberalist priorities, even during the Cold War period. As Kahler writes, interwar liberals "had been absorbed by the issues of war and peace. Neoliberalism in the 1960s and 1970s was drawn to the implications of international economic change" (1997: 33).

In sum, the sundry institutionalists (or "neo-liberals") cannot be criticized for overtly articulating a universalizing and homogenizing agenda. However, if there are no grand schemes revolving around world government or even an explicit defense of global capitalism, neo-liberals do not totally escape some form of cultural imperialism. In particular, the "transactionalism" usually associated with Deutsch *et al.* (1957), but integrally part of the entire interdependence literature, reflects the cultural globalism of liberal thought. Deutsch *et al.* were interested in communicative ability as both an indicator and a necessary condition of integration, and this general emphasis on the multiplicity of channels was to permeate other studies of transnational cooperation among Western societies (e.g. Haas 1958). (Zacher and Matthew [1995: 132–3] place this work within the subdivison of "sociological liberalism.") As neutral as the tone of the argument may be, and without attributing particular motivations to authors, only a fine line separates mere observation of communicative integration from the ethical defense of that observed reality. Integration theory is cultural theory: it follows a path to peace and security according to international exchanges destined to integrate modes of living, with all the dangers for assimilation which this may entail.

Integration theory would (hastily) be pronounced "obsolete" in the 1970s (Haas 1975),[9] in view of repeated common-market failures in the South, but re-emerged under different labels and arguably achieved a synthesis in the writings of Rosenau, where expressions of cultural globalism are rampant. Hints of Rosenau's thought have already been offered, yet a few more details are pertinent here. Rosenau's works have essentially focused on "world politics" as processes within a large (global) polity, where the relationships between political actors necessarily transcend state boundaries and evolve along the historical current of modernization. Rosenau begins with a dynamic ("cascading") concept of interdependence, and, drawing particularly on insights from organizational theory, investigates the learning capabilities of actors as they associate and dissociate in an increasingly complex world. Rosenau's chief interest is precisely in the integrative and disintegrative tendencies of the contemporary world, although his work is scarcely a critique of modernity (see Rosenau 1990: 12–13). His discussion of "sources of change" is, arguably, largely one of symptoms of change, which is very different from a critical perspective. These include such elements as "transnational issues" (pollution, disease, etc.), decreased governmental problem-solving ability, "organizational decentralization" (or "subgroupism"), and the increase in individual skills and self-consciousness. As a liberal, Rosenau embraces modernity, and is more concerned with describing the erosion and assertion of various actors within the modern context. Thus, seeking to demarcate himself from statist theories of IR, he uncovers a "bifurcated" world of states and non-state actors (for which he assigns new labels), and analyzes their respective roles in what amounts to a gigantic, open-ended system (see also Rosenau 1986, 1984).

In line with neo-behaviorist social science, Rosenau painstakingly resists specific value commitments in his scholarship. Yet he is not entirely successful. Behind the "neutral" description of global centralization and decentralization lurk certain assumptions about the good, a good which is presumably associated with a particular remedy for conflict and instability. The clearest normative position is his confidence in expertise (to which we will return). Yet, admittedly, he seems much more careful on the issue of homogenization, selecting as a preferred world a continuation of the bifurcated status quo, balancing evenly the centralizing and decentralizing forces which he sees as most important (namely, states and NGOs); this is a middle-of-the-road position, which Rosenau describes as merely "pragmatic" (Rosenau 1990: 461, 447).

However, this is not where our reading of Rosenau should linger. We rather become interested in the globalizing dimension of Rosenau's cultural message. The reference to culture is indeed deliberate, partly because cultural arguments are vital to liberalism and partly because Rosenau often couches his own arguments in specifically cultural terms. Indeed, in *Turbulence*, Rosenau often invokes a "global culture" or a "culture of world politics." Yet, while Rosenau is right in discerning a certain elite (and mass) convergence toward both rationalism and Western cultural products, he seems to underestimate both the existence and the benefits of cultural diversity and to attribute an unwarranted pacifying power to Western cultural hegemony. As a whole, then, the Rosenauian world is one in which "global culture seems likely to undergo transformations toward a broadened conception of self-interest and an acknowledgement of the legitimacy of interests pursued by others" (Rosenau 1990: 421).

Despite Rosenau's apparent scholarly detachment and positivist pre-theory-building exercises, his work is shaded in liberal optimism and, consequently, the liberal approach to globality. It is one in which the oppressing, totalizing "ideologies" are disappearing (the people of the world may now breathe more easily, empowered as they become with the universal discovery of rationalism and the apparent death of Big Brother); there actually *is* a world political culture (and so the people of the world may finally understand each other, shedding the pettiness of local interests); finally, globalization is occurring largely by itself, with scarce help from the powers controlling the means of communication.

Similarly, the liberal character of much of the peace research and WOMP scholarship is particularly apparent in its approach to globality, usually insisting on some form of (supposedly benign) world government as a path to a better world. Peace is defined anywhere from the absence of war to an approximation of the positive ideal (Boulding 1978: 6; Russett 1982: 173; Gurtov 1988: 50–51). The unifying concept is undoubtedly that of "global humanism," popularized by Robert Johansen (1980: 21–2). Global humanists articulate a value structure aiming at a proper balance between the commonality and the differences in humankind: global institutions are to secure a peace framework based on demilitarization, material well-being, human rights, social justice, and ecological health. Global human "interests" and a global community are both immanent and good (Johansen 1980: 20; 1982: 57). Even Southern contributors to the WOMP share this global outlook, and while they may be more sensitive than Northerners to the cultural imperialism of Western liberal thought, they still favor some form of overarching authority as an essential path to peace. Ali Mazrui, torn between his Africanity and his Western training, seeks solace in an awkward (and ultimately unconvincing) advocacy of a "world federation of cultures" (1976).[10]

The universalizing tendency of liberal IR theory has several implications for ecological thought. The utilitarian perspective would arrive in concurrence here, since the rational use of nature has generally been regarded as synonymous with the western-engineered megaprojects that have been favored by international financial institutions and aid agencies. This has shifted as a result of the problems those projects have created, but they remain the focus of a package-project that emanates from a neoliberal political and economic agenda. Authoritarian perspectives would generally reject liberalism's globalization agenda, since the former focus on community-level cohesion while neoliberalism remains rooted in the freedom of the individual to pursue wealth. Radical ecologists would be the harshest in their denounciation, however, arguing that universalizing theories diminish the diversity essential to an ecological society; and that the liberals overlook key structural constraints and consequences with their individualism and thinly veiled westernizing agenda.

The utilitarian basis of order

Historically, liberal peace has been indissociable from a belief in material progress – in other words, from the domination and use of nature by human beings. A thorough review of works by mainstream trade theorists would quickly become redundant and venture far beyond those works selected here for their location in the genealogy of IR theory. But we should bear in mind the direct relevance free trade and growth have for liberal theorizing, beginning as we noted earlier with Kant and moving into the present.

Open commercial lanes presumably improve the chances for global order by increasing material bounty, directly reinforcing (political) rapports of friendship, dissuading enmity by increasing its "opportunity cost," and, therefore, instilling an element of "stability" in international relations. Of course, ebullient nineteenth and early twentieth century demands for liberal trade reflected a historical context unfamiliar with ecological crises and global social injustice. Cobden's stance is quite transparent: war does not pay, trade does, and trade is brought (and reinforced) by peace. Cobden made points about the fiscal burden of militarization (Cobden 1903: 194) and the counterproductive influence of sea power, unequipped for commercial diplomacy and arousing foreign resentment: "these vile feelings of human nature . . . have been naturally directed . . . to thwart and injure our trade" (ibid.: 229). The general tone demonstrates a genuine desire for peace, as much for its own sake as an instrument to riches: "free trade . . . arms its votaries by its own pacific nature, in that eternal truth – the more any nation traffics abroad upon free and honest principles, the less it will be in danger of wars" (ibid.: 222).

Most of the argument is upheld by Angell. As mentioned above, the core of Angell's thought is based on a recognition of mutual vulnerability in a modern world linked by financial capital: war implies a marked, global reduction in standards of living – a repudiation of progress. It is, in this understanding, a step backward (classical realism would consider it another step in an ongoing circle). The utilitarian approach to peace is apparent in Angell's frequent discussions of the capital costs of war and, in fact, of the actual benefits of selective conquest (Angell 1911: 138). There is an overwhelming focus on material progress as the key to happiness, and Angell provides us with the most strikingly utilitarian statement in our study of liberal international theory (emphasis added):

> Struggle is the law of survival with man, as elsewhere, but it is the struggle of man with the universe, not man with man . . . *The planet is man's prey.* Man's struggle is the struggle of the organism, which is human society, in its adaptation to its environment, the world.
>
> (Angell 1911: 177)

No such radical language is readily gathered from our (admittedly) restricted review of Woodrow Wilson. The latter's decisive commitment to a trading order is nonetheless famous, as reflected in Points 2 and 3 of the January Address: "absolute freedom of navigation upon the seas, outside territorial waters, alike in peace and in war . . . ;" "the removal, so far as possible, of all economic barriers and the establishment of an equality of trade conditions among all the nations consenting to the peace . . ." (Link 1984: 536–7).

The same assumptions are no less essential to the functionalist school. While functionalists concentrate their theorizing effort on the (depoliticizing and self-fulfilling) process of technical cooperation, it is clear that the purpose of cooperation is to ensure the efficient, global delivery of "tangibles" – goods, but especially services. Thus, related objectives of trade and growth are unavoidable here. Even Laski, the social democrat, would write plainly that "a tariff *for revenue only*, as opposed to tariffs which attempt to protect the domestic industries of a given State, seem [sic] to me a clear path to international peace" (Laski 1967 [1925]: 614; emphasis added). And Mitrany (1966: 96) specifically tied international services to a higher human ideal, "contribut[ing] to the achievement of freedom from want and fear . . . broaden[ing] the area of free choice for the common man."

The post-war, non-normative liberal literature would, of course, steadily refrain from uttering such statements. Certainly, not much evidence may be excised from our sources to directly uphold utilitarian arguments; at most, one may assume that the confidence in modern technology, expressed above all by Rosenau, must logically extend to a support for freer trade and innovative means for (mass) production. Still, as a whole, the detached neo-liberal literature has played an important role in perpetuating the argument for a growth-led peace, precisely by not questioning this particular foundation of contemporary "institutionalism." The various regimes analyzed by neo-liberals (mostly in trade, finance, and resource management) are key to the international, growth-oriented order urged by the classical liberals.

More than hints of the argument also seep through the peace and world order literature. The strong academic relationship between peace research and neo-behavioral institutional analysis logically commands a commitment to growth. Bruce Russett (1982: 188, 191) insists on the unprecedented "prosperity" of the modern age, and includes "moderate growth" and a "high level of economic activity" among several necessary conditions for peace. This partiality for growth also characterizes the Southern literature within WOMP, heavily influenced as it is by the South's "inferior" political position and, more pertinently, its incapacity to fulfil the basic needs of many of its people.[11] Growth leads to equity, which leads to peace and stability, as those Southerners accept the liberal competitive credo. Thus Mazrui (1976: 293) writes that "in their [the South's] relations with the developed world the task should remain one of increasing the competitiveness of at least a region as a whole within the southern hemisphere;" and the "reciprocal vulnerability" advocated elsewhere merely restates the assumptions of classical trade theorists and, even, some deterrence theorists! Mazrui's position is far from marginal, encompassing evidently the dependency literature, but also extending to even more radical WOMP scholars, such as Kothari.[12]

In sum, liberal IR theory, as sampled here, draws a near consensus on the possibility of a peaceful order through material growth which, according to the liberal formula, is best attained through open-door policies of trade. The noted exception is Mazzini, who specifically condemned the utilitarian and materialistic approach to social progress. Mazzini is not a marginal footnote in the history of international liberalism, and should not be easily dismissed. His fading legacy as a revolutionary liberal, of tremendous appeal to many Southern intellectuals, rather serves as a reminder of the divisions within liberal thought.

State and technocracy in liberal thought

As much as liberalism is pulled between plurality and homogeneity, it is also torn between equality and hierarchy, between empowerment and depoliticization, between individuation and technicity. The tensions may not be easily abated, in spite of liberal claims. Liberal IR theory has stressed the importance of "expertise" in solving common problems; this is especially so of the epistemic community literature. Liberals indeed assume a fundamental convergence of interests in society, even in a "society of states": the point is to educate parties into "seeing the light" and/or to use rational skills in identifying the location of a mutually acceptable agreement. A problem-solving, liberal order will thus rely on international functional agencies and their technical experts, on international law and its "impartial" authority, and on a global education of masses toward one or the other version of the truth. Environmental problems can best be managed by establishing a context in which there is "a new obligation emerging for governments to take part in a deliberate, pre-programmed process of institutional learning" (Sand 1990: 36).

Law, education and rationality have evolved as basic liberal themes. As hinted above, an essential tension lies between the emancipatory and managerial dimensions of liberal thought. In principle, "managerism" is alien to liberalism, which should presumably reject the centralization, discipline, and relative disempowerment associated with managerial order. A more "realistic" (Grotian?) appraisal of the political environment would seek to maintain the liberating core of the theory, yet develop policies in line with the reality of power and hierarchy in a system of states. While the romantic Mazzini celebrated statehood for the sake of the people, contemporary liberals theorize the state in terms of its interests and its presumed rationality.

The argument linking global stability to functional depoliticization has been suggested several times above. From the early neofunctionalism a flurry of theoretical variants evolved, including Peter Haas' focus on epistemic communities (P. Haas 1990), Oran Young's analysis of "institutional bargaining" (Young 1989a and b), and Ernst Haas' work on learning (E. Haas 1990). Actually, in its pure version, that of Mitrany, depoliticization was meant to transcend the state system which the post-war liberals accepted as given; Mitrany sought a "working democracy" to replace a mere "voting democracy" (Mitrany 1966: 36). Aware of the divisive, parochial, and, indeed, disempowering, influences of states, Mitrany would have them integrate in some form of super-state – yet one not prone to tyranny, but serving as a problem-solving centre, staffed by presumably apolitically appointed experts. What would appear as a formalized separation between state and society would be, in fact, a (liberal) withdrawal of the bureaucratized state at the service of an integrated global society (Mitrany 1966: 92). While Ernst Haas exposed Mitrany's political naiveté, he did emphasize the integrative possibilities inherent in functional organizations (E. Haas 1964: 35).

The (again, indirect) relationship between expertise and conflict resolution is also expressed by Rosenau; in fact, this is the domain where Rosenau's account of process most readily yields to normative statements. Early in *Turbulence*, for instance, in a discussion of the "underlying order" which apparently exists objectively, he states that "human intelligence is capable of resolving or at least ameliorating problems" (Rosenau 1990: 49). But the specific preoccupation with problem-solving ability is more fully conveyed later on, when we read that "[the] frequency and scope [of errors and misjudgments] seem destined to diminish as the microelectronic technologies become standard equipment in foreign offices" (ibid.: 323) and that "human intelligence cannot take full advantage of artificial intelligence" (ibid.: 332). Rosenau may caution repeatedly against technological havens, yet most such statements are immediately followed by an optimistic counter-response.

The legacy of functionalism has even extended to the post-war normative literature. Johansen specifically identifies "depoliticization" as a path to the reduction of war, and, as with Mitrany, links the concept to a focus on global constituencies (Johansen 1982: 57). The role of experts is no less crucial for Johansen, as he advocates a "centralization of functional control and planning" as one of two directions of "power diffusion", along with the "decentralization of political structures" (Johansen 1980: 32–3; see also Mische and Mische 1977). While Johansen seeks to foster a global society with shared values, he holds it contingent on the enforcement of law – the law of states, at the outset, but ultimately the law of the world state.

Evidently, then, the theme of functionalism, conveying the favorable liberal assumption about objective knowledge (and the scientific bias for systemic order), may be logically related to those other themes of law and education: experts guide the legislative process, while the "necessity" for experts elicits widespread training; each element plays a crucial role in maintaining the ordered liberal system.

This said, it would be unfair to depict the entire liberal literature according to the conservative framework enunciated above. The global humanist tradition, particularly, approaches the role of law from the specific perspective of the individual: legislated peace, here, is aimed at ensuring minimal welfare and security conditions for the planetary citizen, rather than merely entrenching the sovereignty of states. This tradition has been inspired by the seminal defense of peace-through-law in the contemporary literature, articulated by Grenville Clark and Louis Sohn. Refining an argument dating back to Wilson and, in fact, surely to Grotius, Clark and Sohn stressed that "there can be no peace without law." However, as mentioned, this law is not necessarily designed to preserve a rigid state system, whose multiple and unaccountable jurisdictions have directly contributed to war and suffering. The emphasis is thus on world law, "uniformly applicable to all nations and all individuals in the world and which would definitely forbid violence or the threat of it as a means for dealing with international disputes" (Clark and Sohn 1958: xi). Is positive law an unambiguous solution to the problem of peace? Proponents of law can scarcely avoid its critics' emphasis on the dependence of law on power – in fact, to the expression of law as power. Surely, the international law of great powers, defended by both liberals and realists, can only promise the peace of the strong – an interstate peace and the global imposition of certain values usually associated with the successful economic system. The implications for environmental policy are immediate. Where one might see a convergence of interest between indigenous groups and multinational pharmaceutical firms pursuing bioprospecting, the interests of the latter are dictated by a western worldview and they offer at best a chance to modify indigenous knowledge with utilitarianism, not the other way around.

Yet what about the world law advocated by the normative liberals? Is there any way that such law could truly reflect a *sui generis* global consensus and be implemented independently from the global power structure? This is not an easy question. Most proposals are essentially based on the United Nations' format, which, in fact, is already structured as a world government, issuing and feebly enforcing "legislation." Those proposals seek to sharpen both enforcement measures and the values typically embraced by the UN (disarmament, economic equity, individual dignity, "sustainable" resource management, etc.). The Clark-Sohn Plan, for instance, lists six "basic principles" and three "supplementary" ones, demonstrating the authors' understanding of the economic, social, political, and military dimensions of a peaceful and stable order.[13] The intention is clearly to remove powers from sovereign states to the benefit of the world authority, and to implement effectively the basic liberal values mentioned above (which are all progressive). Yet, again, can the proposal lead anywhere but to a new form of statism – of "machinery"?

Similar questions may be addressed to the more recent generation of global humanists. Johansen (1980: 31) realizes that a world government is very difficult to implement, but does not seem unfavorable to the idea. He proposes a world "governing machinery," with all the predictable elements: assembly, council, administration, security and economic agencies, human rights commission, environmental authority (ibid.: 32–3). Can this machinery indeed merely "coexist with [a] global populism . . . transcending the limits of class and national boundaries" (ibid.: 35)? Gerald and Patricia Mische (1977: 67) also want to assert the centrality of the individual in global politics, and global legal structures play an indispensable role in that matter. Again here, the description is familiar: assembly, constitution, judiciary, executive, monitoring system, enforcement system, fiscal powers, grievance system (Mische 1982: 76–8). Similar approaches may also be found in Mazrui and Kothari.

Finally, liberal theorists of "world politics" are usually fond of stressing education as a (complementary) path to order. This was a theme favored by Mazzini, who urged to "recognize no privilege except the privilege of high-minded intelligence as designated by the choice of an educated, enlightened citizenry to develop talents and social forces" (Mazzini 1945: 32). The same preoccupation is evident in recent prescriptive scholarship. Gurtov (1988: 172) argued that "education will be a crucial source for promoting global awareness and Global-Humanist values;" Patricia Mische (1982: 75) stressed that "the importance of education cannot be overemphasized;" and Mazrui (1976: 483) wrote that "a world which is governed on the basis of a federated system of cultures has to put a special premium on education and training."

It may be argued that a discussion of education takes us away from the specific dynamics of "international" politics. Admittedly, this is at best a public policy issue, around which there is no readily identifiable problem of (international) collective action; liberal institutionalists do not theorize about education (and would certainly not object to "better education"), and if the issue must be debated, we should perhaps solicit the input of the numerous scholars working on the topic in fields totally different from international politics, though there is some overlap here with those currently embarking on "global policy studies" (see Nagel 1991).

At that rate, however, the already uneasy relationship between the normative and positive literatures in international relations would surely turn into a dialogue of the deaf. More to the point, the theme of education may be legitimately invoked in view of its relationship to science-expertise and law-order. Yet caution is required again, as in the discussion on law, for not all advocacies of "education" are elitist, obsessed with high technology, and otherwise aiming at the solidification of the industrial/capitalist structure. Kothari, for instance, is well aware of the different edges to education as a power tool and as a source of social renewal, and his own advocacy is specifically tailored to the reinsertion of the marginalized individual in his/her community (Kothari 1974: 62–5). However, without disputing the motivations of authors quoted above, not all ambiguity may be shed. Education should not be confused with yet another globalizing attempt, however unintentional, at imposing modernity on (typically Southern) people who actually know better. Global humanists may be given the benefit of the doubt and Mazzini, writing in different times, is understood as such. Education takes its place on the liberal path to world peace, oscillating between the blinding future of modernity and its often sorry past.

An ecological assessment of liberal tenets

Radical ecological thought predictably provides a stinging criticism of this literature. Most ecologists, actually, would examine very closely such a universalist worldview, centered on the individual and exporting the pleasure principle through free-market ideologies. If ecology is to stand for diversity, then it will be suspicious of liberalism as a global cultural equalizer. Liberal IR theory upholds the innovative potential of the rational individual, and, actually, many ecoradicals would agree, for libertarian thought is a powerful current within ecology. We have also mentioned the links between ecosocialism and functionalism, with their emphasis on welfare distribution and technocratic expertise. But the major predilection of liberal thought remains freedom of the individual, of movement, of property, and, ultimately, of thought itself.

When freedom of thought is blindly equated with entrepreneurial freedom, then nature is likely to suffer. And so radical ecologists have been critical of free-trade arguments. While trade can be positive, excessive specialization, dependence on distant suppliers for expanding essentials, uneven terms of trade, and overproduction are all seriously threatening. Bioregions (or ecosystems) may suffer directly, with necessary political impacts, or, conversely, political conflicts may erupt from adverse economic conditions and jeopardize both individual security and surrounding nature. Ecologists would look at the globalizing ideology of trade-based growth as an offered panacea for global problems, a formula that is misleadingly stable (based on presumed economic "laws", fostering regional uniformity) and excessively dynamic (taking "too much" from nature and upsetting local lifestyles too drastically).

Furthermore, many ecologists would be suspicious of liberal reliance on "experts," and even the liberal devolution to law. Authoritarian environmental thought may prefer technocracy, provided it is combined with harsh necessity and, in its more arcane forms, an overarching concept such as ethnic solidarity. Conservationist utilitarians would tend to see expertise, and its functional power, as either a necessary evil or a potential mitigator of market excess. But there the acceptance of technocratic necessity would end. Radical ecologists in particular would find many problems with the idea of expertise in the ecological realm.

First, these are mechanisms which, in liberal IR theory, are partly intended to secure interstate peace – yet the peace of the state is not necessarily that of the individual. Secondly, the reliance on experts, by depoliticizing issues of contention (i.e. by turning to technocracy), would appear as a direct violation of democracy, which most eco-radicals value dearly. Thirdly, radicals will use the time-honored argument that the legal mind-set of liberal theory merely formalizes power relationships: contracts and acts of legislatures are not products of independent ("objective") reasoning, but elaborate attempts at legitimizing the social power of elites who are scarcely concerned with nature and the common good. Finally, while enthusiastic calls for education do project an appealing future of general, elevated wisdom, the same reservation may be expressed: is "education" yet another path to elite control, to ideological control of the masses?

An ecological assessment of liberal IR theory inevitably insists on themes already much discussed by critics of liberalism as a broader ideology and philosophy. But, again, the exercise remains very much valid, for liberal thought is addressed to all nations and to their mutual behavior. Liberals seek global objectives of freedom, progress and peace; ecologists would welcome these, but a radical interpretation of those general terms would likely be incompatible with the utilitarian, technocratic, and homogenizing core of the liberal worldview.

Is liberal IR theory to be thoroughly indicted by ecologists? From a less radical ecological perspective, the friction is actually much less severe. Various eco-authoritarians would seize the technocratic dimension of functionalist thought to their advantage; in a seemingly perverse way, the "liberal" Mitrany reinforces the division between state and society, and offers the same kind of legitimacy to experts as granted by proponents of the Green Leviathan. Eco-utilitarians, for their part, would react most positively to a literature that equates environment with ecology and eminently endorses prudent management of natural resources, with or without recourse to market mechanisms. The most widely read literature on "international environmental affairs" within the field of IR may be found within the liberal stream. It is strictly a process-oriented literature deriving from a mostly American interest in so-called international regimes, exploring the various mechanisms by which state leaders have ensured international cooperation over resource issues; these include epistemic communities (i.e. experts, again), and various forms of interstate bargaining.

As indicated above, the emphasis on transnationalism, proclaimed by the complex interdependence school as a vital addition to realist foundations, leads to a universalizing premise. It is also, as our discussion of Angell suggests, not as new as it appears. As Fred Halliday reflects:

> There is in much of the transnationalist literature an element of historical foreshortening. Many of the processes – economic, political and religious – that characterise contemporary transnationalism were present, if not to the same degree, decades and even centuries ago . . . There is so much . . . underlying optimism and teleology in liberal internationalist writing that as a body of work it is eerily reminiscent of an earlier generation of literature on the transition to socialism: "setbacks" and "lags" there may be, but in the end it is all bound to happen.
> (Halliday 1994: 105–6)

What is new, we are often told, is the sudden emergence of *the environment* as an issue-area worthy of similar proclamations of newness.

The realization that problems of the global commons require multilateral solutions or management systems is seen as a stimulant towards Kantian federation, or even Mitranian functionalism.

Eco-radicals would not see anything new in such literature, largely because they believe IR theorists on "the environment," as social scientists interested in description, do not question the very bases of modern society and, consequently, neglect the truly emancipatory potential of ecological thought. As social scientists, liberal IR theorists are obviously aligned with realists in their commitment to a positivist approach to knowledge, and, as explained earlier, this "epistemology of rule" (Bookchin 1991: 89) is anathema to radical ecology.

Finally, we should refer back to what is arguably the most contemporary and popular version of neoliberal IR theory, the spirited push for globalization defined as the emergence of a single global marketplace and, in some of the even crasser variants, a single global culture. Assuming this spirit is not crushed by other events, such as the summer 1998 collapse of the Asian markets, it would seem as though many governments are willing and able to adopt globalization as their mantra for the next millennium. Liberalism is the ideology for globalization (or perhaps globalization is the ideology for liberalism – the two are almost indistinguishable when considered internationally). With the exception of utilitarians who equate progress with the spread of Western technology, there are few ecological thinkers who will not view globalization with a mixture of concern and even disdain. We expand on this in Chapter 5.

CONCLUSION

Liberal IR theory has deep roots derived from the same intellectual paths leading to the utilitarian environmental perspective prevalent in Western society. This is one reason why, it would seem to us, it is mistaken to consider realism "hegemonic" in IR theory: only a perspective on that body of work which completely ignores the environment, or reduces it to elements in geopolitical strategy, would come to such a conclusion. In particular, the current debate over whether globalization exists, as structure or process, lends weight to the importance of liberal thought on freedom (personal and commercial) that realism's state-centricity virtually ignores. Further, any review of contemporary regime literature on the environment (Sand 1990;

Haas, Keohane and Levy 1993; Rittberger and Mayer 1993; Brenton 1994; Caldwell 1996; etc.) indicates a strong bias toward liberal precepts related to the value of contractual relations and institutional regulation.

The obvious resonance with utilitarian ecophilosophy need not be belabored at this point. A liberal world economy will need relatively open access to resources, and nature plays this role with the often disastrous consequences we have seen emerge in the last decade. Eco-radicals would tend to view the current managerial impulse that guides the neoliberal institutionalist regime literature as an effort to clean up the mess caused by the underlying utilitarian philosophy.

We have argued liberal IR theory has universalizing tendencies, and presents a prescription for order and peace that is contingent on free trade on the one hand and technocratic managerialism on the other. How the latter two strains will work themselves out within liberal IR theory remains to be seen (they may be simply mutually reinforcing in the long run) but it is doubtful they will encourage the type of deep structural and normative thought eco-radicals argue is necessary to overcome ecological crises in a sustainable manner. To eco-radicals, globalization is more of the same. Some liberal theorists, falling into what has been termed the *reflectivist* or *constructivist* schools, feel institutional interaction changes the interests of participants (Kratochwil 1989; Wendt 1992). This accent on norms, as opposed to process, is promising; but it hardly challenges the ecophilosophic roots of common problems, and may be viewed as a recent manifestation or perhaps sophistication of Mitrany's central institutional thrust, much as the same can be said for the epistemic community framework.

Authoritarian approaches would be most critical of the role liberals place in liberty; the right of the individual to pursue personal wealth is at the root of the problem of the commons and, barring the development of a mutually-beneficial regime, the Green Leviathan becomes necessary to supersede these particularistic ambitions. In fact, the liberal assumptions about rationality are viewed with equal suspicion by authoritarian and radical ecologists. And, of course, liberalism's optimism and Western orientation are causes for concern for realists and critical theorists, repectively. It is to the latter that we turn next, in our continued search for locating ecological thought within IR theory.

Chapter 5

Critical IR theory and ecology

INTRODUCTION

We turn now to the task of situating ecological thought within the many critical waves that have swept over IR theory. In order to do so we make the relatively non-controversial assumption that realist and liberal thought, as described and evaluated in the previous two chapters, constitute the "mainstream" of IR theory. Although many branches of contemporary critical approaches have roots in Marxist literature which predates most of the IR theory to which students are routinely introduced, much of it emerged in the 1970s as a refutation of the centrality of realism and liberalism. Critical approaches target the assumptions on which the mainstream has been constructed, as well as the epistemological (positivist) foundations of the sub-discipline of IR theory as a whole. One might argue that the question of epistemology is the central concern here; critical theorists argue that positivism limits the ability of IR theory to move beyond descriptive analysis and into normative work with an active, "emancipatory" agenda.[1]

It should come as little surprise that delineating critical IR theory is not necessarily an easy task, particularly since many social theorists reserve the term "critical" to applications of the Frankfurt School of critical theory. Critical theory is often understood as anything remotely attached to Marx or other leftward icons; though dependency theory and, indeed, Marxism is central to an understanding of IR theory today, we resist the urge to equate all forms of critical theory with Marxist analysis. We will approach critical IR theory from a larger spectrum, including both foundationalist and antifoundationalist strands in the literature, and also including purely descriptive attempts at understanding international governance. We cast a wide net not merely in recognition of those original efforts at reconceptualizing IR, but particularly because much of that literature may claim some relationship to ecological thought, even if the issue of ecology is rarely articulated therein. In fact, we find that such disparate strands of critical IR theory converge, to some extent, from an ecological perspective.

The critical literature encompasses reflections on both norms and process. Normative writings stress the need to put IR theory to good social use, i.e. to articulate/revise conceptions of the good within IR theory, particularly towards the goal of emancipation. This desire to improve the world has often guided those within the critical stream, and the identification of values that sustain harmful practices is given explicit attention. Those who focus on process argue for a more differentiated understanding of the global dynamics shaping politics, moving beyond state-centricity, rational-actor models, and regime dynamics. The two critical strands come together often, since they both are apt to emphasize the importance of a holistic approach, one that is not limited by the constraints of established compartmentalized social science, and one that tends toward the "global." We see an immediate resonance with the radical ecology literature, which also stresses the need for holism and global thinking.

In fact, there would appear to be many paths worth exploring between both ethical and descriptive critical IR theory and the ecological thinking outlined in Chapter 2. For example, the normative (emancipatory) standpoint of radical ecology relates to that of the WOMP, dependency theory and feminism; the constituencies vary, but the critiques of hierarchy and positivism converge. At a broader level, it is the holism of critical IR that will most likely appeal to a broad spectrum of ecological thinkers, dissatisfied as they may be with the reductionism of mainstream IR. However, we must recognize the near-total absence of ecological reflections within critical IR; beyond usual references to "environmental issues" as triggers for a reconsideration of IR precepts and policies, there seems to be limited awareness of the significance of ecology for a critical theory of IR. The possible exception to this emanates from the subfield of international political economy, where scholars are beginning to apply the insights of theorists such as Karl Marx, Karl Polanyi, Fernand Braudel, and Robert Cox to an ecological approach to globalization (see for example Saurin 1996).

IR THEORY AND GLOBAL PROCESS

Criticisms of established IR theory have often emphasized the statism, or state-centricity, of the discipline. International politics has been constructed largely from a narrow understanding of international processes as a function of interstate behavior. The liberal literature on integration and interdependence sought to uncover some of the international dynamics related to non-state actors, in particular multinational corporations and non-governmental organizations. Yet the subsequent regime literature confirmed that the popular image of IR inescapably pertains to patterns of harmony, cooperation, and discord (Keohane 1984) amongst state representatives. Obviously, MNCs and NGOs have increasing influence and play at least mild causal roles in the regime literature, but intergovernmental negotiations and the resultant institutional impacts are the main focal point. Structural realism, meanwhile, emphasizes continuity of process: the permanent struggle for survival of similar units. Regimes are mere epiphenomena, and sub-state actors are largely excluded from the parsimonious model altogether.

IR theory has long faced the necessity to better incorporate local and national dynamics into its *problématique*. Most theorists would recognize that the traditional division between comparative politics and international politics is needlessly restrictive, and that an astute understanding of politics should recognize the fluidity of political processes (Strange [ed.] 1984; Milner 1992; Caporaso 1997). This can only be done by broadening the scope of the analytic exercise. As IR researchers, we are not merely interested in interstate behavior, but in the politics of conflict and cooperation. The task, then, is to "globalize" the analysis of politics. This imperative recognizes the similarity of political processes within and between states; anarchy is not the purview of IR, as contemporary concerns with "failed states" reminds us, and hierarchy is far from absent at the global level. In fact, both realism and liberalism rely on anarchy and hierarchy simultaneously, but do not delve into the global repercussions of these ontological referents. The desire to globalize the study of political process also recognizes the formidable communicative and cultural convergence between societies across the planet on the eve of a new millennium.

Moreover, a globalized view of politics presumably extends beyond the problem-solving literature that has delineated the field in the twentieth century. In other words, understanding politics should presumably account for the normative (or cultural) underpinnings of politics that have evolved through history and guide decision-making on a regular basis. From this perspective, violence between states and violence in the street are reflective of a similar political dynamic. International trade and domestic commercial competition would be similarly related; and the causes and human consequences of low-income urban centers in Los Angeles and those in Sao Paolo would more than overlap. This is what should attract ecologists, as theorists concerned with global interrelations. The stress on a broader conception of interconnection, on a global scale, unites the globalist literature and distinguishes it from mainstream IR theory which prescribes limited interconnective analysis.

Several versions of "globalism," most of them convergent, compete for attention in IR theory. Rosenau's has been discussed, and we identified its liberal bent. Not surprisingly, then, Rosenau's name rarely appears in the new debate surrounding critical theory in IR, as nowhere in Rosenau's language may the reader discern a rejection of positivism. As explained earlier, Rosenau still pursues "objective" descriptions of reality (though he might resist the criticism), arguing that a "conception of simultaneous globalizing and localizing dynamics ... will emerge as the basis for the prime descriptor of the collective ontology that replaces the Cold War" (1997: 230). This may be so, but falls short of what most critical theorists would consider an emancipatory or epistemologically challenging stance. However, this is not to deny Rosenau's contribution to multidisciplinarity in IR, for his imaginative work still reminds the reader that world politics is in constant flux and thus cannot be captured neatly. This can be contrasted to the effort at parsimony offered by neorealism, or even models of hegemonic stability such as that offered by Gilpin.

The same message is echoed in other globalist writings. Structurationism was popularized by sociologist Anthony Giddens (1984) and formally introduced to IR in the last decade (Wendt 1987; Wendt and Duvall 1989; Ruggie 1989; see also Robertson 1990 and Bergesen 1990).[2] As a sociological theory, structurationism seeks an explanation for the (re)production of social institutions. The approach is eminently dialectical and holistic, resisting an explanation of behavior based on the ontologically reducible "actor" or "system." Structuration suggests a process by which agents and structures ontologically co-evolve, mutually shaping one another through the routine activities of daily life. The structurationist does not believe in the possibility of identifying a "first cause," since social "events" can only exist in an extensible "locale" of time-space. By definition, a dialectical ontology of agents and structures compels a reappraisal of time and space in political thought, neither of which can now comfortably support specific delineations.[3]

Structurationism may well obliterate all distinctions between IR and comparative politics, leaving in its place the study of a single social system reproducing globalized institutions and practises. Wendt does try to salvage the field by describing states as social entities, constituted by social structures of either "domestic" or "international" dimension, and of either "economic" or "political" character (Wendt 1987: 366). Whether such dichotomies, along with Wendt's insistence on a scientific approach to social change, can effectively capture the holistic and postpositivist character of structurationism is highly debatable (George and Campbell 1990: 277 n15). Yet there is little doubt that structurationism remains a suggestive approach in understanding a "global politics" which the liberals had barely begun to explain with the concept of transnational relations; and relevant case studies may be generated in the future. Note, however, Steven Weber's main methodological complaint: with structuration theory and the reflectivist approach to the study of international institutions, "there remain too many potential closer circles, too many possible self-fulfilling stories among which the approach cannot clearly distinguish" (Weber 1997: 244).

The basic structurationist idea of (global) institutional reproduction at the local level has been shared, in fact, by other theorists. Recalling the Marxian roots of dialecticism and historicism, we can point, without surprise, at dependency theory as the initial contemporary contribution to a critique of IR theory. Dependency theory is a neo-Marxian analysis of "underdevelopment" in the South; it is largely the work of radical economists in Latin America, with specific debts to American and French Marxists. Dependency theorists describe what they see as the (systematic) process of Southern marginalization in a global capitalist system controlled by Northern agents and Southern elite collaborators. Put in other terms, development and underdevelopment are two results of a universal process of capital accumulation; underdevelopment should not be considered as the original condition in an evolutionary process of "modernization" (Blomstrom and Hettne 1984). Ecologists will appreciate in dependency theory the work of normatively inclined sociologists and economists in understanding global processes of domination.

Among other points, dependency theorists have examined the phenomenon of the enclave economy, which incorporates key ecological factors, such as resource extraction and the commodification of nature (in this case, largely for export).

While the substantive focus of dependency theory may be debated, both its ontology and epistemology can be readily appreciated by the critical globalist. Cardoso and Faletto's "historical-structural method" "emphasizes not just the structural conditioning of social life, but also the historical transformation of structures by conflict, social movements, and class struggles" (1979: x). Thus, echoing Cardoso's earlier dismissal of positivist attempts at appropriating dependency theory (1977), they insist that "the basic methodological steps in dialectical analyses require an effort to specify each new situation in the search for differences and diversity, and to relate them to the old forms of dependency" (Cardoso and Faletto 1979: xii-xiii). Likewise, Samir Amin decries the economism of social science thinking:

> The very search for unilateral causalities between "independent variables" and "dependent variables" is characteristic of mechanistic economism and is diametrically opposed to the dialectical method where the whole, i.e. the reproduction of the conditions of the mode of production, determines the parts, i.e., the "variables."
>
> (Amin 1977: 185)

Some differences do remain between dependency theory and the theory of structuration. The latter is much less concerned with normative issues and, especially, much more agnostic as to the outcome of the evolving global system; Giddens' world, according to David Jary (1991: 141), is of "competing social movements and competing nation-states as well as a world of capitalism, with no predictable outcomes." The individual motivations of structurationists may well tend towards various forms of emancipation, yet their discourse on process is articulated above all in a detached and scientifically propitious manner; this is not to portray dependency theory as a non-rigorous stream of literature, but rather to emphasize its political role and, as well, the importance it continues to attach to the liberatory mission of unionized labor and state forces. However, it is more important to stress the convergence of dependency theory and structuration than their differences. From an ecological perspective, the key is their common rejection of functionalism and structuralism, though the same ecological perspective would benefit from a more systematic debate concerning the locus of power and change in global society. Still, it is worth remembering how the Marxian analysis of capital played a key role in clarifying the hidden process of domination and historical change at the global level; and it is the same Marxian influence, through Gramsci, that brings us to Robert Cox and to his contribution to globalism.

Cox is now a fixture in the discipline, and at a basic level, one might argue that his insistence on asking who the theory is written for, and not just who it is written by, is the major contribution to the development of a critical strand within the field of IR theory. Cox's work indeed provides the principal reference point of many reviews of critical IR (e.g. Hoffmann 1987; Linklater 1990; and especially Gill and Mittelman 1997). Just as "critical theory" is usually understood as the post-Marxian attack of the Frankfurt School against instrumental rationality, Cox's Gramscian analysis of world order surely constitutes a "critical" turn in IR theory, and can be interpreted as contributing to the field in a number of ways. For example, as mentioned in Chapter 3, in a recent essay Richard Falk (1997) includes Cox with E.H. Carr and Hedley Bull as the principal "critical realists" in the field of IR theory.[4] Falk argues:

> [a] vital part of Cox's originality and significance as a "realist" is to conceive of power in Machiavellian/Gramscian terms that stress the influence of ideas and cultural primacy rather than to conceive of "reality" by exclusive reference to the equations and hierarchies of brute force. Such an intellectual framework of analysis shifts the emphasis from stability and continuity to change and discontinuity.
>
> (Falk 1997: 43)

Cox clearly established his epistemological position in his landmark article of 1981, by dividing the field into "critical" and "problem-solving" approaches and arguing, as he would a few years later (Cox 1989: 37), that modern theories of political science (process-oriented) may focus either on the decision machinery or on the (necessary historical) path leading to the sheer creation of deciding agents. In other words, theories are interested either in contractual or pre-contractual dynamics: they either explore the processes leading to political agreements or disagreements at a specific time and a specific context, or investigate the multifaceted historical developments leading to that context and that specific political confrontation. The latter exercise, which Cox favors, presumably offers a deeper understanding of the motivations of political actors, and thus may be particularly well positioned to evaluate the possibilities for change in political order. In sum, one could say that contractual theories dissect the precise workings of the political "machine," while "historicist" theories explain how and why that "machine" was initially constructed.

In his major work (1987) Cox delves into the historical roots of American hegemony, using Gramsci's model of the evolving historic bloc: modifying Marx, and partly echoing the *dependentistas*, it locates capitalist hegemony in a (global) structural alliance between state elites, monopoly capital and science. The key to this historical argument about hegemony is undoubtedly its cultural element, eminently emphasized by Gramsci, and most directly responsible for the routinized and relatively unforced acceptance of order. Political legitimacy and historical change, then, become a function of culturally-grounded (historical) structures. Cox and Giddens, the Gramscian and the structurationist, speak here with one voice: Cox defines historical structures as "the cumulative result of innumerable often-repeated actions" revealed intersubjectively, and thus rendered "objective independently of individual wills" (Cox 1989: 38) – an argument strikingly similar to Giddens' discussion of "practical consciousness." Cox stands as a major critical theorist of IR precisely in view of his empirical work, deliberately linked to a broader theoretical argument about hidden power structures and routinized behavior in the global economic arena.

Critical theory, however, has had a few other contributors worthy of mention, whose source of inspiration derives from French postmodernism. The work of Foucault, for instance, has become influential. Admittedly, there are fundamental differences between the rationalist, modernist positions of the Frankfurt School and the deconstructive project of postmodernism. However, both are equally critical of positivism, and both, again, are concerned with "elusive" processes of social control and institutionalization. In each we may find alternative approaches to power and historical change extending to the relationship between knowledge and power.[5] More than any other postmodernist thinker, Foucault has contributed to the broad discipline of political science. He developed the concepts of "mentality," "power/knowledge" and "pastoral power," which effectively describe the subtle hold of elites on potentially deviant masses, namely through the means of social science research and the spread of cultural exemplars.

These Foucauldian themes demand a globalized reading of politics and a (re)examination of "power fields" serving the existing market-based industrial order. In the field of IR, James Keeley provided the first noteworthy attempt at applying Foucauldian concepts, in this case to an analysis of international regimes (Keeley 1990a). Keeley's objective was avowedly limited, introducing the concept of power/knowledge and pointing at the relationship between regime maintenance and the "sharing" of discourse. Yet his essay, arguably still unrecognized, may be considered pathbreaking, particularly in light of the popularity of the regime theory which he has consistently criticized (Keeley 1990b). Interestingly, as well, Keeley does see his argument converging with Gramscian notions of hegemony (1990a: 92–3), attesting to a general wave in twentieth-century critical scholarship against the traditional conception, and construction, of "reality." Similarly, Karen Litfin has written a well-received "discursive analysis," based on Foucauldian precepts, of the negotiations surrounding the creation of an ozone layer protection regime. The stress on the power of scientists as actors whose power "derives from their socially accepted competence as interpreters of reality" (Litfin 1994: 29) is particularly appropriate given the mainstream epistemic community literature and the move toward technocratic managerial responses to environmental problems resulting from globalization.

The postmodern current, for all its insistence on textuality and deconstruction, can set itself constructively in this diverse attack on positivism. Postmodernists may not be empiricists, but their language is not necessarily incompatible with a new type of empirical work geared toward uncovering global processes of (dis)order. Richard Ashley notes:

> The poststructuralist wants to know what is repeated, what structures and practices reappear in dispersed sites, and how these replications can be accounted for . . . (S)he wants to speak of effects [and] wants to understand the workings of power in the most general terms, and she wants to understand power's relationship to knowledge.
>
> (Ashley 1989a: 278–9)

As James Der Derian (1997) and other poststructuralists have pointed out, their work is not "inherently antiempirical." Overall, however, postmodernism's contribution to IR theory essentially lies in its emancipatory critique rather than in its discussion of process, which remains at a very high level of generality.

THE NORMATIVE CRITIQUE OF IR THEORY

A sizable portion of the critical IR literature is dedicated to turning IR theory into an instrument of social change – or, at the very least, to demonstrate how the conservative underpinnings of IR theory has long impeded its role as an instrument for change. More precisely, these theorists point out how mainstream IR theory has continually emphasized the description of some ineluctable political "reality" – a "reality" that is not only fixed, but that is detrimental to many groups and classes on the planet. As positivist social science popular with state (and other) elites, mainstream IR theory has frozen in time a "reality" that is highly hierarchical and conducive to militarization and exploitation. Shouldn't IR theory serve some emancipatory purpose, in view of the continuing marginalization of women, the Southern poor and non-human nature? Shouldn't IR theory recognize that, by serving the gods of instrumental rationality, it has contributed to the refutation of Enlightenment ideals? Shouldn't IR theory appreciate (or at least investigate) its influence on the hegemonizing current of Western culture and institutions? These are the questions underlying critical IR theory, as it moves beyond a reconceptualization of process to a discussion of ethics.

Clearly, there is ecological significance to a body of theory committed to freedom, using theory as a means to freedom. Just as clearly, ecological thought, even in its radical expression, is torn between two seemingly contradictory norms: its insistence on biological and cultural diversity, and its (scientific, naturalist) defense of biological essentials. In social terms, ecological thought oscillates between foundationalist and antifoundationalist arguments, i.e. between logics of universality and particularity. In other words, just as it views suspiciously academic and political attempts at grounding a "truth" in presumably universal "foundations" (i.e. unshakable assumptions about human nature or culture), it cannot accept a form of relativism that would offer no tangible guidelines in constructing a better world. In sum, how does ecological thought compare to critical IR's anti-hegemonic challenge? We investigate critical IR theory from this standpoint.

The postmodernist critique of IR theory, particularly that of R. B. J. Walker and Richard Ashley, is well-known for its antifoundationalist stance. Walker's argument runs through his many writings, and is particularly apparent in his rescue of Machiavelli from realist clutches: "Machiavelli struggled . . . to speak about *lo stato* against a discursive hegemony of scholastic universals" (Walker 1993: 47). Ashley states the antifoundationalist point very directly: "The task of poststructuralist theory is not to impose a general interpretation [. . . It] eschews grand designs, transcendental grounds, or universal projects of humankind" (ibid.: 284); "one must be prepared to give up the time-honored dream that theory, in constructing knowledge, can plant its feet in some absolute foundation . . . beyond history and independent of politics" (Ashley 1989b: 286). A hegemonic version of reality thus must be avoided, "as if all people everywhere," Ashley cynically adds, "would necessarily agree as to what their real dangers are" (Ashley 1989b: 287).

The feminist literature also offers its contribution to an emancipatory critique of IR theory.[6] The main lines of criticism are well known and are adequately summarized by Tickner (1992), who notes that not only is the language of IR theory sexist, but it offers a masculinized reading of world history, ignoring the role of women. International Relations indeed appears as the male bastion *par excellence*: its tradition is heavily influenced by military studies, legitimizing assumptions about incessant conflict and power quests, at ease with the description of hard-nosed ("calculated") negotiations, and endorsing the sexually demeaning language of war (Cohn 1987); the field of "cooperation" may seem more tolerant, *a priori*, but rational-choice approaches or an analytic emphasis on legal-institutional aspects do not convey a radically different ("less male") reading of the world. In sum, feminists argue International Relations marginalizes the role of women in historical development and favors an ontology and an epistemology constructed by men.

Feminism itself is as diverse a body as ecological thought, and also displays conservative and progressive extremes. The feminism of concern to us here is not the "liberal" type, which radical feminists argue merely co-opts women into the power elite and into male rationality. Feminist critics of IR theory are more appropriately located within a critical stream whose main interest, to quote Jean

Elshtain (1987: 258), is to "deviriliz(e) discourse," not in favor of a "feminization" which would perpetuate gender duality, but towards a political awareness of hegemony in its many forms. Spike Peterson has argued recently that the masculine-feminine dichotomy was:

> The key to *longue durée* of a (Western) mentality celebrating human (read elite male) agency, control and certainty – a mentality associated in the modern era with (masculinist) science, capitalism, and instrumental technologies, and in the present moment, with a crisis in ways of thinking/knowing.
> (Peterson 1997: 191)

In sum, the feminist critique of International Relations does not necessarily purport to satisfy social-scientific (positivist) demands for a rigid "research programme" yielding "testable propositions" and "verifiable truths". On the one hand, as hinted above, there is some ground for empirical research, explored notably by Cynthia Enloe and Carol Cohn, and focused on the exclusionary practice and language of theorists and officials alike. Yet a key result of feminist scholarship is its demonstration of how a discipline, by its sexism, reifies itself and sustains particular ontologies and epistemologies with devastating impact on "the weak." Feminism may not yet be able to explain how inter- (and intra-) community relations may be reconstructed so as to effectively transcend gender duality, and so as to solve the dilemma between commonality and difference. But at the very least, it has turned the promotion of a specific constituency into a wider critique of foundationalism in one of the least flexible fields of social studies.

The unity-diversity dilemma has also been examined from a "classic" constructionist perspective by a no less "critical" theorist of IR, Andrew Linklater (1990 [1982] and 1990). Linklater stands in partial contrast to Ashley and Walker: he pursues an emancipatory framework for humanity based on a recovery of the cosmopolitan ideal. As a critical theorist, he understands the historical contingency of rationality and appreciates the contemporary exhaustion of the state, and of the idea of citizenship, as a rationalist solution to the dilemma. His liberal argument (and ultimate goal), that "moral development involves the progressive universalization of norms" (Linklater 1990 [1982]: 211), does reflect an avowed foundationalist concern (ibid.: xi) and will raise suspicions from postmodernists and many ecologists. Indeed, the contemporary global crises invoked by the author in his discussion of sovereignty are not theorized and, therefore, eschew a discussion of the very universalism from which they have emanated. However, at the same time, and as much as Linklater's call for emancipation-in-order (echoing Giddens) may sound rebarbative (Linklater 1990: 31–3), he deliberately steps beyond the bounds of mainstream IR theory, entrusting the individual human being with the necessary power toward freedom. In the final analysis, Linklater may well sound unconvincing, for much as IR theory made "citizens" out of "men," his attempt at replacing "man" amongst his global peers raises more questions than it answers. Yet his discussion has played (and is still playing) a vital role in supporting a key development in IR theory: the increasing attention devoted to new social movements as *bona fide* actors in the global process and as potential agents for radical, progressive change (ibid.: 26). Environmentalism, in its various forms, must be considered a key agent in this context.

Indeed, the relationship between social movements, social change, and social theory is at the heart of most understandings of critical political theory. In IR, of course, the enterprise dates back at least to Marx, and was revived by the *dependentistas*. However, for all the positive aspects of dependency theory outlined above, its emancipatory mission is hampered in two ways. First, it remains committed to a modernist (and materialist) ethic of economic development. This limits its potential as bridge to radical ecological thought. Many dependency analysts are intricately tied to Southern (mostly Latin American) reformist projects. While they may articulate a "basic needs" discourse, they are motivated primarily by the desire to equalize economic and international state power. This does not deprecate their commitment to the Southern poor and/or marginalized, but emphasizes the inevitable limits to advocating Northern modernity for the entire planet. Second, and as a corollary, the emphasis on middle- and lower-class economic justice does not carry the analyst to the depths of critical theory articulated elsewhere. Dependency theory does remain associated with the "old" social movement, i.e. labor, and has not really widened its analytical scope beyond narrow economic relationships.[7]

The case is different with Cox. His central interest in the relationship between productive forces, state formation and world order does follow the Marxian tradition, yet his approach does not convey the normative overtones associated with Marxism; in this sense, Cox's work remains essentially focused on process alone. Yet Cox also opens more doors for emancipatory change, perhaps because he recognizes the divisiveness of class-based approaches to the good life. His historical approach appreciates the continuing cooptation of income-based groups in the globalized economy and the potential alternative offered by new social movements (Cox 1989: 48). In sum, the emancipatory potential of Cox's theory is a function of his flexible historical approach, his insistence on uncovering the economic dimension of order (thus his reluctance to reify either "the state" or "political man"), and his clear shunning of a materialist ethic. Cox does not have a blueprint for a better world, but he attempts to explain why humankind has arrived at a historical threshold and why the hope for revolutionary, broad-based social action may not merely constitute wishful thinking.

AN ECOLOGICAL ASSESSMENT OF CRITICAL IR THEORY

We have argued that the explicit attempt to set an agenda for a "post-positivist" IR is what links diverse strands of critical theory. Are they linked further with an appreciation of the ecological questions raised in Chapter 2? Rather than token reference to environmental problems as new and transnational, and indicative of broader social crises and contradictions, has ecology made more headway here than it has in the mainstream?

Before steering entirely clear of the mainstream, however, we might briefly investigate subtle ecological intrusions within the more reformist literature which seeks a reappraisal of realist and liberal thought in IR. Our main reference here is to the policy-oriented literature on world order and peace research, usually excluded from the narrow confines of IR theory. Again, the WOMP writings of the 1960s and 1970s did not wholly succeed in shedding a questionable universalist ethos and in eschewing the growth models which they were debunking. Northerners within WOMP appeared overly confident about the benign character of "global humanism," while it could also be argued that the Southerners' legitimate quest for global economic equality compelled them to advocate dangerous growth policies in their home countries (see Kothari 1974: 69).

But it would be inappropriate to dismiss the world order literature or peace research, at this stage. The WOMP, specifically, evolved from its legal-functionalism of the 1960s to a more elaborate discourse, in the 1970s and 1980s, focused on the rights of the oppressed and buttressed by extensive empirical study of the global system. This evolution is succinctly documented by Richard Falk and Samuel Kim (1982), two central figures of the Project. While the ecological problematique would be tackled more seriously only in the late 1980s, there are many early references to ecological balance as part of a more general normative framework for a peaceful future, especially evident in Falk's pathbreaking *This Endangered Planet*, which not only popularized the idea of ecological crisis but challenged the sovereign state system in the process (though it might be argued that such early challenges merely reinforced a realist ontology).[8]

Indeed, ecology is essential to the crucial concept of positive peace, popularized by Johan Galtung. Among other WOMP-type scholars, it receives attention from Mazrui:

> The value of ecology is still *derivative*, linked too directly to the needs of man ... There may still be one more step to take. The step does imply going back to totemism, and investing in the environment a value independent of man.
> (Mazrui 1976: 45; emphasis in original)

Gurtov (1988: 50–51), for his part, articulates an ecological awareness throughout his well-known book, relating it specifically to positive peace. Johansen (1980: 20) lists "ecological balance" as one of four key global problems, "also ... stated as world order values." Patricia Mische (1989: 416) moves beyond the conventional literature by advocating a cultural basis for a global ecological peace. Kothari, for all the contradictions inherent in his policy proposals, is perhaps the most explicit regarding an understanding of the relationship between ecology, peace, freedom, and local self-management (Kothari 1974: 10–11). Readily acknowledging his debt to Gandhi, he states his opposition to "the incipient consumerism and the growing giganticism of both the state and the modern economy" (ibid.: xxi).

The "depth" of ecology seems more apparent within Southerners' contributions, as may be further observed in the work of Indian ecofeminist Vandana Shiva, who surely qualifies as an IR theorist, if one considers her critique of the process by which Western capital and science have allied with Southern elites to flood Southern lands with dangerous products (particularly in agriculture).[9] With its open criticism of modernity and patriarchy, her critique is substantially different from dependency theory. In her study of the Green Revolution in the Punjab (1989), she demonstrated how apparent political conflict between distinct nations was, in fact, stimulated by the skewed landholding patterns emanating from high-yield monocultures – as rich Hindus, exploiting foreign markets, gradually took control of the land and expelled the dispossessed Sikh farmers. She noted how monocultures increase local dependence on Western know-how and capital, and how global market forces necessarily create a wage gap amongst farmers. She drew the evident link between generic engineering, "improved strains," and centralized research – seldom to the benefit of local people (see also Shiva 1991: 47). And, of course, she testified to the ecological absurdity of monocultures and non-leguminous, water-hungry cash crops such as rice or wheat.

Shiva thus emerges as one of the few writers stressing the relationship between Western global hegemony, ecocide, and internecine violence amongst the poor. As a critic of modernity, Shiva attempts to demystify the notion of "the global" as a heavily-laden term, popularized by the North, and denying the existence of "the local" – best embodied in Southern rural life: she argues (Shiva 1992: 26) that, through globality, the North exists in the South, but not conversely. Beyond this, what resonates from the writings of many Southern critical theorists who delve into environmental issues, and this is demonstrated in the collections compiled by Sachs (1993) and Brecher *et al.* (1993), is an emphatic normative statement: the world is unjust, ecological degradation is at least partly a function of that fact, and it is the responsibility of Northerners and Southerners alike to take remedial action. Naturally this falls at odds with mainstream IR theory, in particular realism, which eschews such lofty ambitions and settles instead for the imperative of state survival.

Is there any disagreement with the argument that world-order theorists, whatever their academic affiliation, "do" IR? In other words, that they reflect on the processes underlying the global reality and on the goals which communities should pursue in the midst of globality? We would assume not, and so would duly suggest that critical forms of ecological thought have already entered the realm of critical IR theory. Yet should we necessarily label all critical IR theorists as (at least) unwitting radical ecologists? In other words, can ecology supply (and systematize) insights on process and norms discussed by critical IR? And, if so, which ecological thought is at stake?

Answers to these questions require recalling the dual position of ecology vis-a-vis debates on foundation, or universality. All ecologists, irrespective of political orientation, will stress the holistic (thus very complex) process of ordering in nature – this is ecology as science, i.e. as a paragon of universality. Consider, then, how ecological is the message of structurationism. As a purely analytical framework,[10] structurationism's non-linear description of political process evokes the complexity of natural processes and thus may serve as an explanation of ecological degradation – as a discrete, repetitive and globalizing process of change in the modern era. Giddens (1991: 209–10) writes that "one of the key features of modern institutions . . . is that they 'disembed' social relations from local contexts of action;" this is indeed a process of globalization, reflecting "dialectical ties between the global and local" (see also Campanella 1990). Through this interaction between individual agents and global(ized) structures, a routinization of ecological abuse not only takes place, but even becomes acceptable.

Similar parallels may be established with other strands of critical IR theory, which share the ontological and epistemological holism encouraged by ecological thinking. The ecological limits of neo-Marxist analysis notwithstanding, and as demonstrated above, dependency theory is expressly constructed as a refutation of the positivist epistemology pervading mainstream social science. Cox's historicism similarly aims at achieving an understanding of contemporary political affairs as the distinct product of social forces acting dialectically in a historically contingent setting; radical ecologists share such an approach, sensitive as it is to the complexity of social reality and to the hegemonic mission of historic blocs (with decided implications for the control of nature). Finally, much the same may be said in regard to feminist approaches of global process and critiques of male epistemology.

Beyond process, it is the normative mission of critical IR that may appeal to – or may be transcended by – ecological thinkers. Again, the most constructive parallels are to be found with the radical branch, which acts as a critique of Enlightenment ideals and as a focal point for anti-hegemonic currents. The global dynamics that may be responsible for sexism, racism, and various forms of elitism have long been assessed by radical ecologists and their intellectual predecessors, and have pointed to a general, diverse attack on life –

led by related forces within capital, the state and knowledge centers. Radical ecologists are obviously sympathetic to both foundationalist and antifoundationalist attempts at rescuing the marginalized from the clutches of modernity, and can show how all may be subsumed under a wider logic of resistance against what we might term "antibiotism." In this sense neo-Marxian, neo-Gramscian and feminist normative critiques are all presumably ecological.

This said, some currents of critical IR theory would seem at odds with ecology's naturalism, apparent particularly in the ecological mainstream but not necessarily less evident in radical strands. This is the foundationalist dimension of Ecology that would presumably be resisted by postmodern critics, as may be witnessed from Ashley's rejection of Thoreau (Ashley 1989b: 287). Yet even postmodernists should appreciate the message of cultural diversity emanating from radical ecology, one that may still coexist with a recognition of biotic imperatives.

In light of this emerging, yet not fully coherent assault on the mainstream, we have called upon ecological thought as a means to focus the discussion. Explicitly ecological approaches to critical IR remain rare. Beyond the very limited use of the ecological metaphor by the Sprouts (1965), Dennis Pirages (1983) was the first to show the potential of re-thinking IR along ecological lines. Recently, in a lucid display of philosophical research which he combined with work on Brazilian ecopolitics, Thom Kuehls (1996) sought inspiration from Nietzsche and Foucault and explored the potential of a so-called "rhizomatic" approach to world politics; rhizomes are intricate, subsurface vegetation growths that spread horizontally, providing a useful analogy for the limits on political sovereignty posed by global ecology. But this is more metaphoric hypertrophy than the integration of ecological thought in IR theory. Similarly, the regime literature may benefit from a discursive, or even outrightly Foucauldian, approach, such as that offered by Litfin. But this is a methodological re-evaluation of mainstream conclusions about process, not an attempt to move into the dialogue between ecology and IR theory which we have argued here is ultimately necessary.

There have been several ways in which environmental problems have come to play a large role in theory formation. Robert Boardman (1997) has provided a cogent overview of some of them, including the work of the Sprouts and the regime literature, with its emphasis on compliance. He argues that "while particular subfields of IR will continue to push forward the study of international environmental politics, the more sustained development of world society theorising forms the optimal basis for longer-term development" (ibid: 42); world society theorizing refers to approaches which:

> decentre the analytical and prescriptive situation of the state, suspend assumptions about conventional hierarchies of values and issues in IR, open borders to debates in related areas, particularly in social and political theory, and include normative discourse as a foundational concern of the discipline.
> (Boardman 1997: 42, n38)

Identifying the structural implications of ecological degradation emerges as a gateway to interdisciplinary theorizing, and the critical contribution is essential. In our next chapter we will explore this further.

RADICAL ECOLOGY AS CRITICAL IR THEORY?

Though we have found many epistemological and, even, thematic links between ecological thought and IR theory, they are often implicit and elusive. Critical IR theory – with its close affinity to radical ecology – offers a partial exception. While different strands of realism incorporate the environment, most noticeably in the geopolitical tradition, they make little if any effort to incorporate philosophies of the environment in any meaningful way. Indeed, Chapter 3 indicated that the basic ontological and epistemological tenets that propel realist thinking are in essence anti-ecological, with the complex exception of the link between classic realism and authoritarian ecology. Liberalism, on the other hand, is firmly premised on the assumptions held by the utilitarian school of ecology, and has produced a wealth of empirical literature dealing with institutional (managerial) responses to environmental problems. Yet to argue that contemporary liberal IR theory (even if we include the realo-liberal complex interdependence school in this broadest of categories) is profoundly influenced by ecological thought would be to confuse causality with commonality. It may derive partly from utilitarian assumptions regarding nature, but it has hardly been transformed by the realization that those assumptions have caused ecological harm.

Only in the radical/critical schools of both sub-disciplines can we identify a symbiotic congruence. Radical ecologists, regardless of their stripe, argue there is a profound crisis at hand today, one that transcends technical solutions and calls for serious reorientations in political order, toward decentralization and local identity and a constant awareness of our essential dependence on nature. Critical IR theory is premised on a similar conviction, though the emphasis on the recovery of nature is usually but one of a series of concerns based on empirical indications that large-scale or deep change is needed. What further links the two is a critical awareness of epistemological assumptions and a flexible, postpositivist inclination, divorcing them from mainstream (and overwhelmingly positivist) utilitarian ecology and IR theory. We might go further, however, suggesting radical ecology itself can be seen as an additional critical perspective within the subfield of IR theory.

Of course, we are not arguing here that radical ecologists have stumbled upon a superior understanding of the international system. As our treatment of various branches of IR theory suggests, we have never set out to declare which perspective has the monopoly on truth. While some IR theorists have certainly believed they have defined the international system, in as parsimonious or complex or even turbulent a manner as possible, it is doubtful anyone would argue that theory will exactly coincide with reality, even if the former has strong positivist inclinations. By and large, radical ecologists would be even less likely to argue they have unlocked the keys to the political universe. Further, radical (and other) ecologists, we have argued throughout this text, have generally declined the task of treating international politics seriously. It too often appears to be assumed that, once the proper ecological "society" (the geographic and demographic and legal limits of which are rarely defined) is created, the global political system will simply sort itself out. Sovereignty, territoriality, interstate competition: all these remnants of the Westaphalian order will simply wither or fall into harmonious place along with the other ordering principles (sustainability, non-hierarchical societies, decentralization, participatory democracy, etc.) of a new ecological society.

The dearth of material pertaining to IR is apparent throughout the ecological spectrum. Ophuls, who might be classified in the authoritarian school of ecological thought, touches only briefly on the matter: his treatment of scarcity extends his pessimism to North–South relations and he calls, predictably, for "some form of planetary government" for global survival (Ophuls 1977: 214). Dobson's acclaimed review of green political thought contains but a short passage on the complications caused by international politics (Dobson 1995: 121).

Well-known radicals, such as Bookchin and Kirkpatrick Sale, fare little better. According to Bookchin, post-state municipalities would "join with other munipalities in integrating [their] resources into a regional confederal system" (Bookchin 1989: 194). This reduces international relations to intermunicipal relations, and in fact Bookchin's emphasis on democratic participation in this context brings him closer to the liberal "democracies don't fight" thesis than he may wish. He also writes, "No municipality would be so far from another that it would not be within reasonable walking distance from its neighbours" (ibid.: 195); so we may presume that intercontinental relations would be problematic! The bioregionalist Sale (1974, 1985) would limit most political interaction to an ecosystem-sized area (Sale ibid.), though how this could be affected with the interdependent and dependent linkages present today is another question altogether. Ecofeminists, for their part, stress the role of social movements in affecting widespread change, but rarely engage in sustained analysis of the transnational implications. Finally, the deep ecologist Eckersley criticizes Bookchin's confederal model and summarily concludes by defending an "enabling state" with a foreign-affairs mandate in re-establishing ecological health (Eckersley 1992: 182), but the prospect of sustained relations between such states remains relatively underexplored.

Rather than offer a blueprint for international order, then, radical ecology offers an incisive critique of mainstream theorizing about international order. This is no small offering, and this book has demonstrated the importance of such an academic contribution. With its explicitly normative focus, radical ecology might be incommensurable with the positivist mainstream, but then again one may discern (as we have attempted to do in this text) assumptions held by realism and liberalism which do impact upon the mainstream's conceptualization of environmental problems that are, in themselves, of normative origin. While it is obvious that radical ecology will find the most friends within the critical school of IR theory, it can also challenge critical theorists who habitually mention the environment as a *cause célèbre* without delving into the deeper political implications of ecological crises. The critique would ask several key questions, many of which we have raised in the previous chapters:

- What are the ontological and epistemological referents employed by IR theories, and how do they define a conception of nature and human-nature relations?
- Do IR theories produce centralizing or decentralizing tendencies?
- Can IR theories explicitly acknowledge the idea of limitations to nature? Is there an ethos of sustainability?
- Does IR theory draw links between the exploitation of nature and human groups, including women and racial minorities?
- Do IR theories reproduce the urge to universalize the Western development experience, one which has led to environmental problems of today? In other words, are they rooted in a positive appraisal of modernity?

All in all, radical ecologists would argue that mainstream IR theory defines international process and norms in ways that fundamentally contradict the principle of sustainability. The mainstream views on process tend to ignore the reciprocal patterns of causation that characterize both physical nature and the global political world, while mainstream IR's emphasis on circumscribed, time-bound and space-bound events bypasses the holistic character of events. Linearity and ahistoricity allow for a much more efficient and manageable production of verifiable hypotheses about the world, but arguably sacrifice to an unacceptable extent some necessary dialectical and historical depth. The radical ecologists' insistence on ecological holism compels them to disregard social-scientific elegance, to avoid simplifying dynamics that are indeed exceedingly complex.

Concerning norms, it would appear obvious to radical ecologists that mainstream IR fundamentally misunderstands the key values which have traditionally motivated IR theorists, as political theorists concerned with the "large scale." IR theory evolved as an attempt to understand the sources of political disorder in a regional or global state of nature: the search for some sort of security, or peace, has since driven the field of IR, even as theorists later explored more specific dynamics of "conflict" or "cooperation". In this context, radical ecologists can remind IR theorists of the limits of a "security" or an "order" that ignores its ecological foundation.

Radical ecology's inclusion within critical IR theory thus can be defended in its support for a more refined approach to process and in its criticism of IR theory as not delivering in its intended normative "promise" – which is to seek a better world for (presumably) all or most of its constituents. Critical IR theorists have used specific empirical examples to underscore their points: the "security" of the nation seems to exclude women, the global poor, indigenous peoples, and other marginalized groups or classes; this is perhaps the case because IR is an elitist enterprise that embraces science as a "power language" and is content with observing relations involving the traditional locus of political power, i.e. the state. Radical ecologists would agree with all this, for they all believe that a sustainable world hinges on the rehabilitation of the marginalized and a more incisive account of the forces that "matter" to political order – forces which include the state, but as part of a larger elite network that is presumably working against the interests of the global majority.

In this light, one may indeed argue that radical ecological thought can provide an original angle to the critique of IR theory. As mentioned earlier, the appeal of radical ecology derives from the balance between its essentialist and relativist arguments, and also from the base which drives the arguments and unites all cultures and societies – namely, the natural habitat. Ecologists will in many cases adhere to a naturalist worldview and investigate paths toward sustainability (which must include the polis). Thus, they will often derive an ethic from nature – seen, in this case, not as conflictual and hierarchical, but as cooperative and "circular." This latter point is key to radical ecology: a Hobbesian or Darwinian survival of the fittest presumably misrepresents the processes which have actually allowed for the continued multiplicity (and not rarification) of species, through time. A naturalist ethic will also insist that certain basic rules of nature must be ensured to secure the long-term survival of nature as a whole: recognizing its exhaustibility, its intricate links, its need for diversity, and its slow process of maturation. Still, if there are rules to be followed, if there is indeed some foundation from which can emanate a truth, the political and epistemological call will emphasize moderation and a global "civic virtue." The lessons from nature seem clear (see Laferrière 1996: 63–5):

– projects are welcome – indeed inevitable – but in moderate size: nature is malleable and invites creativity, but always in harmony with the ecosystem; as much as radical ecologists would disapprove of nuclear power plants and eight-lane highways, they will also disapprove of arguments claiming that global security will be served by the cumbersome machinery and size of the modern state, emphasizing the alienation and stratification that traditionally accompany excessive size;

– centralized decision-making is only efficient in the short-run and for narrow tasks: as no human agent can possibly manage (i.e. control for) the complexity of an ecological whole, no technocrat can possibly manage a human world defined by its sheer diversity; nature does not welcome top-down rigidity and precision, and at some point, order will emanate from below; in other words, decentralized political structures offer better prospects for long-term security;

– monocultures do not exist in nature, and to engineer them politically is to invite dangerous epidemics: hence globalization cannot mean stability, nor can some social contract presumably serving a "one will" or a "general" will; however, extreme individualism cannot stand as a logical corollary of political and social diversity, as nature demonstrates how species survive and thrive according to an elaborate set of social relations;

– nature is a living being that can only be fully experienced (and thus understood) through symbiosis; and so this is essentially Einstein's, Planck's, Bohr's, and Heisenberg's response to Newton, and their message to IR theorists via radical ecology (Jones 1987, 1988): by seeking an objective (and linear) scrutiny of the human machine at isolated points in time, social science serves as a means of social control for the elite, and thus jeopardizes here again the long-term vitality (thus the security) of the political body.

In sum, radical ecology duly recognizes the various critiques of IR theory which have insisted on IR's restricted view of the global political process and contribution to the social entrapment of "weak" constituencies. The field of IR should look at the "big picture," clarify its normative commitments, and act in the interests of the oppressed – this is theory as praxis. In this light, radical ecologists may well seek to argue that critical IR theorists should, at some point, come to ecology to find some common roots, for freedom begins with an appreciation of the importance of natural habitat.

Radical ecologists do not argue that nature cannot be "touched" by humankind, but that only a proper understanding of its relevance (to us) and intricacy (in and of itself) can ensure the normative goal of critical theorists. That goal is usually understood as the salvaging of Enlightenment ideals, but without its scientific and technological arrogance – a constructive project, based on some foundation, but not imposed by parochial interests. We would surmise that even postmodernists, wary of universalist doctrines, would recognize as much.

CONCLUSION

Challenges to mainstream IR theory are now profuse, and they seek, arguably, to undermine the very fabric of the discipline – to blend IR theory into a wider theory of politics that would be more cognizant of global process and more aware of the ethical shortcomings associated with positivist epistemology. Critical IR seeks to describe the political world as a product of historical forces shaped dialectically (and so pays tribute to Marx), yet it also aims at deconstructing mainstream language so as to show the contribution of IR discourse to a project of domination – guided by political, financial and intellectual elites. The attack on the mainstream may thus be witnessed in numerous books and articles upholding a postpositivist (or antipositivist)[11] approach to knowledge, challenging fundamental conceptions of key IR themes (e.g. sovereignty), and exploring the ethical roots and possibilities of the discipline. A conscious effort to shift the focus of the discipline is supplemented with yet another attempt to bring ethical considerations, and in particular questions concerning social justice, to the forefront.

It is easily apparent that the rise of environmental concern in the late 1960s and 1970s had an impact on academia, and that it added to a critical perspective on the ills of modern society. In some cases we might argue that environmental disasters were instrumental in prodding the mainstream toward an acceptance of the issue-area, but that critical theorists had long incorporated this into their thinking. The environmental problems associated with rapid industrialization remain central to a broader critique of that process. However, this does not imply that any of the forms of ecological thought discussed in Chapter 2 have directly influenced the development of critical IR theory. Indeed, few have sought to deliberately "ecologize" IR within the critical wave. Yet we argue that many aspects of critical IR theory may be subsumed within the radical strand of ecological thought. Radical ecologists are often aware of the limits of sovereignty, about the dangers of problem-solving epistemologies, about the global replication of institutions and practices that discourage sustainable lifestyles. They would seem well-placed, then, to offer a framework that would allow for alternative theorizing on the classic questions posed by IR theorists. In the next, and last, chapter of this book, we summarize our main findings, and look in turn at how IR theory can be used as a critique of ecological thought. In this manner we move closer to that elusive goal of conceptual synthesis.

Chapter 6

Conclusion
Maintaining a reluctant dialogue

INTRODUCTION

Cross-disciplinarity, interdisciplinarity, multidisciplinarity: these would seem to be increasingly familiar categorizations in academia today. Gradually, it has been recognized that the older, more narrowly defined fields of social sciences and the arts have been limited by their very definition, and that we need to enhance and broaden the scope of interested scholars. The field of international relations has, perhaps, been at the forefront of this cognitive process, since it has – much like its elder companion, political science – always been based on accepting the contributory value of other fields, such as economics, geography, and sociology.

Much in the same manner, ecophilosophy has necessarily been influenced by more conventional understandings of, among others, philosophy, sociology, psychology, and ecology. And yet we have seen that the crossing of these two fields, IR theory and ecological thought, needs to be coaxed out of the literature with an analysis such as that offered in this text. It is rarely explicit, and whatever congruence we have unearthed here has not come easily. Similarly, while we have argued that critical IR theory can and does make a contribution to IR theory, there is disappointingly little direct interaction between sustained thinking about the ecological implications of human society and the development of an international system.

At the same time, there is increased realization that, in this age of so-called globalization, IR theory cannot be legitimately divorced from more general reflections about political order, and that an understanding of political order, in turn, requires fresh contributions from social theorists of all kinds. Sociologists, feminists, and literary critics (and these categories are obviously not mutually exclusive) have all widened the scope of IR theory. Now, more than ever, would seem an opportune time to add ecological thought to the list of outside influences; further, one might argue, the urgency and potential severity of ecological crises demand this influence be internalized, such that environmental considerations are always taken into some form of account when theorizing about IR.

This last point is no doubt a tough sell in a discipline which has characteristically treated the environment as an issue-area and not an integral component of theory. Yet when we look more closely at the political issues ecology raises we see an immediate convergence in the broad focus of inquiry. A political treatment of ecology raises perennial questions concerning centralization and decentralization; decision-making structures impacting resource use; the ontological and epistemological premises that define our understanding of nature and, thus, our socially constructed assessment of institutions designed to "manage" it; and even broader reflections on order and sustainability. One can argue, and we hope this book contributes, ultimately, to just such an argument, that these are at least some of the major concerns every branch of IR theory needs to consider if it is to take itself seriously; that, without such self-reflection, theory cannot move forward; and that, as argued previously, it is radical ecological thought which most clearly insists on this reflective exercise.

This chapter will reinforce the above generalizations by summarizing the findings of the book and suggesting further avenues for theoretical exploration. If we struggle to find ready acceptance of what we feel is a necessary dialogue between IR theory and ecological thought, we should explain this conceptual reluctance. This explanation may in turn lead us to more productive, evaluative, work.

SUMMARY OF THE BOOK

This book has covered a wide array of topics, and it would be perhaps counter-productive to attempt to summarize its contents in this short space. However, we can delineate the more important elements that have, so far, composed the text.

After introducing its general intent in Chapter 1, we moved on to a necessarily cursory examination of what we have determined to be the three main perspectives in IR theory today: realism, liberalism, and critical theory. Realism focuses our attention on state interaction within an anarchical political structure; its main ontology is one of atomized conflict, and its epistemological root is positivism, though this is complicated by various associated perceptions of human nature. Liberalism stresses the role of the individual, but still relies on a state-based political world; it is premised on the omnipresent possibility of rational co-operation amongst self-interested units, and has led scholars to focus on the empirical attributes and causal roles of international institutions. Critical theory has various branches, from marxism to feminism to postmodernism: what links them is a common rejection of the mainstream.

Our central intention was exploratory: we wanted to take these theoretical perspectives and expose them to the light of ecological thought. In order to do this, however, it was necessary to provide a fairly comprehensive overview of the latter, and this we attempted in Chapter 2. There, we divided a rich conceptual universe into three essential sections, each with several subsections. The first school of thought, arguably the most prevalent today, was Utilitarianism; this perspective considers that nature exists primarily for human use and that sustainability is, in turn, primarily a matter of management (or, to use a stronger term, engineering). We suggested the most common environmentalist strain of utilitarianism is popularly known as conservationism.

Next, we examined a branch of ecological thought that places a premium on authoritarian order in the quest for sustainability. This ranges from the neo-Malthusian work of authors such as Garrett Hardin to the "eco-fascist" literature, which often has pronounced nihilistic and misanthropic tendencies. Finally we introduced the most clustered branch, radical ecological thought, suggesting this could be productively broken down into social ecology, deep ecology, ecosocialism, and ecofeminism. Each of these challenges mainstream environmentalism, indicating the latter misses some point about the connection between environmental degradation and hierarchy. Though there are distinct ideal solutions offered, the common link is an expressed determination to (re)value nature as an integral element of our own survival with intrinsic worth.

At that point it became possible to delve into the three central perspectives of IR theory from an ecological viewpoint. In the cases of realism and liberalism we placed particular emphasis on the philosophical origins of the perspective; since critical theory is a more recent addition to the literature, its discussion focused on more contemporary work. In all cases we were able to unearth certain convergences with ecological thought, but the extent of this varied.

Stated differently, we argue that the key streams of IR theory can all be subjected to an ecological reading, since they all, in some ways, articulate positions which are either conducive or detrimental to the health of nature. For instance realism, as the essential locus of a discourse associated with so-called "high politics," is rarely linked to ecology. However, there is no doubt that, depending on one's ecological perspective, realism is either an enemy of nature or conducive to its protection. Radical ecologists denounce realism's conflictual ontology, its static view of history, its endorsement of national hierarchies and cultural homogeneity, and its positivist epistemology. Utilitarian ecologists, on the other hand, will prefer to stress that realists are prudent and rational, two attributes which will secure the proper handling of natural resources, even in the midst of an arms race. It is this same rationality and, futhermore, realism's call for decisive state action which will find a favorable audience among those authoritarian ecologists seeking quick fixes to scarcity. Again, realists need not discuss environmental issues to be relevant to the environment: what matters is their assumptions about nature and their preferred path to order.

Liberalism may be equally scrutinized by ecologists. Liberal IR theorists infrequently discuss ecology, yet their worldview does have ecological significance – again here, for better and for worse. While the positivists' influence on liberal IR theory often limits the latter to scientific exercises in political analysis, liberal IR cannot shed the various assumptions about nature and humanity which have long underpinned liberal political philosophy. Liberals have faith in human progress, rationality, and self-regulatory ability; they also seem to believe in the capacity of nature to sustain an ever-expanding global economy. Thus, one can easily associate liberalism with a utilitarian ecological perspective, sharing an appeal to universalism and technocracy, and a belief in the essential freedom of the individual to pursue material gain. Liberalism, as we noted in Chapter 3, is pulled between plurality and homogeneity, equality and hierarchy, empowerment and depoliticization, and individuation and technicity; mainstream ecological thought suffers the same fate. Both eco-authoritarians and eco-radicals would see liberal assumptions as anathema to ecological sustainability; radicals would specifically resist liberalism's lack of genuine concern for limits to growth. Neoliberal institutionalism appears as an effort to manage, if not conceal, the damage caused by this carelessness.

We noted, finally, that references to ecology in critical IR theory are almost as sparse, but that the objective and language of critical IR is quite congenial to radical ecological thought. The critical literature challenges the mainstream at its epistemological root, as radical ecology does to utilitarianism. Its global perspective on international political process emphasizes intricate patterns of causation – an approach sensible, actually, to all ecologists. It is, more clearly, critical IR's normative commitments which specifically appeal to radical ecologists. By questioning mainstream IR's hegemonic discourse, critical IR has defended social theorizing as a practical means to freedom for various constituencies – which, for radical ecologists, should include nature as a whole. Thus, in Chapter 5, we suggested that radical ecological thought should be considered a viable, if emergent, strand of critical IR theory. The questions radical ecologists raise are fundamental to a nuanced understanding of the international system that goes beyond the realist and liberal models. Arguably, an ethic of care, tolerance and cooperation can be derived from nature, and presumably be applied to international relations – even if we are a long way from its realization.

As we approach the end of this book, we should briefly address two other questions. First, how is mainstream IR likely to respond to an ecological assessment or critique of its key axioms, descriptive foundations and policy prescriptions? We obviously believe that ecological thought does shed light on the claims which mainstream IR holds for the international system, yet can the mainstream find arguments to the contrary – and effectively dismiss yet another intrusion on its paradigmatic territory? Second, assuming that a productive dialogue may exist between the fields of IR and ecology, what future avenues for research may be envisaged?

ECOLOGY AND MAINSTREAM IR: ASSIMILATION OR TWO SOLITUDES?

Our discussion in Chapter 5 showed some possibilities for an interdisciplinary rapprochement between ecological thought and IR theory. Obviously, juxtaposing these two fields makes eminent sense. After all, various IR theorists deal with environmental issues, which obviously bind states and societies; "respecting the environment," "creating an ecological society" cannot be a purely autarkic exercise in a technologically advanced world.

Ecologists are moved by the suggestion that environmental degradation is the clearest reflection and the most important cause of social disorder. Ecology is literally the "science of the house." The house is our natural habitat; it is burning, and so the political community is in danger. Thus the purpose of ecological thought is to understand why our house is being ravaged, and to find ways to extinguish the blaze.

Mainstream IR theorists, for their part, are driven by the recognition that armed conflicts are always possible in a world of states, yet that avenues for political cooperation nonetheless abound in that world. The world is fractured, history is marked by evidence of (international) political breakdown, yet international political order is not an unattainable goal. Thus the purpose of IR theory has been to understand the various dynamics that may lead to contractual collapse and success, and to suggest mechanisms for longer-lasting (international) political order.

In sum, one could legitimately connect the work of ecologists and IR theorists by recognizing their driving interest in political order (a broad concept that surely includes those of peace, security, and stability). Yet, presumably, they will emphasize different dynamics of order, and will perhaps criticize one another for missing the key elements – and, therefore, for contributing to a status quo or proposing an alternative that is dangerous to society. Criticisms could pertain to the respective definitions of order (based on various readings of human nature and history – ontology and teleology), and to the way in which knowledge of the world is gained (epistemology). The key would obviously be the relevance of ecological considerations in determining order for the political community. In other words, do the workings of ecosystems offer clues as to how the political community must be constructed? Do ecosystem dynamics instruct us on a definition of causality, essentiality and purpose in human affairs? "Misunderstanding" the latter will surely lead to "misguided" recommendations for political order.

Though ecologists rarely discuss international relations explicitly, it would be harsh to argue that their reading of the world and their prescriptions necessarily bypass the reality of the interstate system. For ecosocialists, for example, it would seem sensible to assume their awareness of the global character of capitalism, and hence of the globalized pressure on the planet's ecology. Few ecologists would need to dwell on the relatively mundane point that pollution crosses political frontiers and poses problems of collective (state) action, or that environmentally dangerous practises are now part of a global economic and technological culture. Ecologists of all kinds will surely recognize that militarization is a usual byproduct of international rivalries, and that protectionist interests may delay international environmental agreements. Further, those ecologists trained in various forms of critical thought will surely point out that the war system is also fuelled by narrow interests within corporations and bureaucracies, and that rules for global free trade and investment are likely to endanger national environmental legislation; countless books and articles of unspecified (yet implicit) ecopolitical orientation have stated as much (e.g. Morris 1990; Finger 1991; Ritchie 1992; Thomas 1995).

And so one could perhaps argue that politically aware ecologists have relatively little to learn from established theories of international relations, although IR theorists would probably beg to differ. Can mainstream IR theory in any way inform ecologists on the sources of and solutions to political disorder? A vindication of the status of IR theory vis-a-vis ecological thought would presumably assume the following: first, that ecoradicals have scarcely advanced propositions which mainstream IR has not yet refuted; second, that the arguments of other ecologists, particularly from utilitarian and Hobbesian perspectives, can readily be assimilated within established IR theories.

Consider, then, the several questions which realists would pose to many ecoradicals. For instance:

- Is human nature necessarily cooperative?
- Is capital accumulation really the driving force behind environmental problems and general political disorder?
- In fact, does the so-called ecological crisis even exist?
- Also, is there any value to the utopian thought characterizing Bookchin's work? Can one really assume that order can emerge and be sustained from below?
- Is there any practical value to considerations of nature as other than use-value – since utilitarian valuing of nature is the approach favored by elites?
- Furthermore, why should IR theorists resist a scientific approach to political behavior? Is there anything wrong in assuming

recurrence in human affairs, identifying social facts and isolating causal links between variables? Why would such epistemology be responsible for environmental problems?

Liberals could voice their own concerns, some of which are shared by realists. They may ask why some ecoradicals are so ready to denounce the reliance on technical experts as they seek to solve social problems; a technically complex society requires expertise, and experts are presumably at the service of society. Alternatively, liberals may question why such ecologists are so propitious to see cultural imperialism in attempts at creating a global market. Besides, why do so many ecologists deemphasize the palpable risks posed by global cultural diversity for political order?

As for the relationship between IR theory and other strands of ecological thought, one may very well argue that utilitarian and authoritarian ecology cannot easily provide a critique of IR theory. Whatever contributions made by those schools to the issue of (international/global) political order may be subsumed under existing frameworks – which have held disciplinary ground for a long time. Conservationism, for example, is undeveloped as a political theory, and can be seen as part of the contractualist, liberal wave. As a political programme, it implicitly endorses the legitimacy of states and corporations as political actors contributing to political order.

Conservationism's direct relevance is in policy advice, warning established actors to moderate resource use and employ science and technology to extract maximum (sustainable) yield from nature. It does not theorize on the international or global sources of political disorder, other than implying that the various political actors actively cooperate so as to implement necessary conservationist legislation and institute proper corporate behavior – but always in moderation. Conservationists see nature the way that most IR theorists would see nature – as use-value, as environment, as multiple issues to be managed independently (for purposes of efficiency and assuming that incremental success is a step to general success). Conservationists thus have a theoretical and political affinity with liberal IR theorists. To state this is to clarify what "ecology" undergirds liberal IR theory, and to show that ecological thought can be readily assimilated within the utilitarian-liberal framework of mainstream IR.

Green-Leviathan approaches, on the other hand, remain Hobbesian approaches to political order, and while they may be informing policy makers of a hitherto unsuspected source of disorder (i.e. natural-resource scarcity writ-large), they hardly contribute a fresh solution to disorder. This branch of ecological thought vindicates realism to the extent that realists appreciate the importance of a strong state in establishing political order. Yet this is not to say that eco-Hobbesians have an elaborate theory of IR, realist or otherwise.

As for the other ecological fringes, none would seem positioned to challenge the status of IR theory or to otherwise sustain some fruitful dialogue. Ecofascists are part of an idealist trend stressing the "cleansing" character of human death, and so offer evidence to realists that irrationality has to be rationally countered; as we know, much of realist thought is intricately tied to the nineteenth-century romanticism that would spur the rise of the extreme right. Finally, misanthropic Gaians are related to ecofascists by a nihilistic bond. As anti-moderns, and so as part of a movement insisting on the mystical character of nature, they may raise questions concerning the prevailing view of nature in IR theory. Yet they have shown little theoretical depth as political thinkers, and so remain marginal to our discussion.

CONCLUSION: SOME THOUGHTS ON FUTURE RESEARCH

While we played "devil's advocate" in the preceding section and explored some arguments limiting the dialogue between ecology and IR theory, we steadfastly hold to our belief that the two fields have much to learn from one another. Quite possibly, one may argue that IR theorists should disproportionately benefit from this exercise, since they have defended for so long particular values of order, security, peace and freedom which can only be realized fully within ecological constraints (which they have ignored). However, we have also noted how ecologists have tended to theorize the ecological society at the local or, at most, regional level – in other words, extrapolating global order from local schemes of sustainability.

Therefore, as an initial suggestion, it would seem necessary for ecologists to directly tackle the global problematique at the prescriptive level. Descriptions of global pressures on ecological degradation already abound (particularly in analyses of MNC activity and international economic agreements), and so the key here is indeed prescription. Is the state a viable actor for an ecologist? How can Eckersley's enabling state or Ophuls' "ecophilosopher-kings" effectively avoid the corruption and tyranny long associated with state structures? Conversely, if Bookchin claims (as we read him) that an ecological society based on direct democracy may be implemented following a reform of the existing state system (and not the violent collapse of the latter – which is just as possible), how clearly does he delineate this evolution? An answer to this question is essential so as to evaluate the feasibility of his confederal municipalism, for direct democracy is incompatible with the high-technology environments which are not going to disappear by themselves.

Turning the table around, how may ecological thought, and particularly radical ecology, further probe into IR theory or otherwise alter its configuration? On the one hand, we must point out that the review of IR theory and theorists performed here remains embryonic; there are many more authors that may be scrutinized, within and outside of the mainstream, in this century and before. Furthermore, one could deepen the analysis of certain key themes within the field of IR. One of us, for example, studied the specific concept of peace within IR theory from a radical ecological perspective (Laferrière 1995), and many innovative works could presumably emerge from an eco-critique of, perhaps, freedom or sovereignty in IR theory; sketches on the latter theme have, in fact, already appeared (Litfin 1997).

Finally, there may be interesting possibilities for empirical research, particularly in the analysis of the environmental discourse of key political actors. What is the "exact" meaning of an environmental clause to a free-trade agreement? How ecological is a global declaration or treaty protecting indigenous rights? How does a department of defense rationalize new equipment purchases as increasing national security when their ecological costs are shown to be immense (thus very detrimental to public security)? These are questions which various commentators, in the press and academia, have already posed in many ways, yet never from a larger theoretical context which would unify ecological thought and IR theory.

In sum, we hope to have demonstrated the grounds for a more sustained dialogue between IR theory and ecological thought. These are two fields of theory which share common objectives: that of describing processes of political disorder and of suggesting solutions for political order. Therefore, there is much common conceptual ground to explore and policy-oriented work to be performed. With due time, we dare believe that this emergent dialogue may evolve into a more systematic interdisciplinary synthesis.

Notes

1 Introduction: unearthing theoretical convergence

1 *Agenda 21* was the main package of suggested priorities and policy recommendations that emerged from the United Nations Conference on Environment and Development, Rio de Janeiro, June 1992.
2 As indicated, evidence for this general proposition is overwhelming. See for example the many *Worldwatch Papers* and the *State of the World* series published by the Worldwatch Institute; hundreds of reports by the United Nations Environment Programme (UNEP) and other UN bodies; government reports, and intergovernmental reports such as the Intergovernmental Panel on Climate Change (Leith *et al.* [eds], 1995).
3 *The problématique* is conceptualized as

> The problem of all the problems, not merely the sum of the problems of pollution, war, famine, alienation, resource depletion, urban crowding, and exploitation of the Third World by the First. It is a systemic construct that assumes causal connections among these problems, connections that amplify the disturbance in the metasystem.
>
> (Ernst Haas 1983: 39)

An earlier work that raised this issue was Falk (1972).
4 The term hegemony is used here to refer to the Gramscian idea of a dominant societal perspective, and not the wider use of the term by international relations theorists who are really referring to the power of state entities, in particular of course the American one.
5 Problem-solving theory requires a synchronic approach: it "takes the present as given and reasons about how to deal with particular problems within the existing order of things." The second is, on the other hand, much more akin to what historians often label a diachronic approach. In order for us to generate critical theory, we must stand "back from the existing order of things to question how that order came into being, how it may be changing, and how that change may be influenced or channelled" (Cox 1994: 101).

6 The language of early realism in particular is exclusive of women; it is unclear whether women share this innate desire for power or whether it just doesn't matter if they do because they are assumed to be powerless in the inter-state world.
7 Aron (1966: 73) also believed, however, that this seemingly structural logic could not be divorced from the "intoxication of ruling" that absorbed state leaders.
8 Thus Robert Gilpin's now-famous and blunt line: "the fundamental nature of international relations has not changed over the millennia ... a struggle for wealth/power among interdependent actors in a state of anarchy" (Gilpin 1981: 7). Gilpin's discussion of hegemonic transition relies on this ontology.
9 Of course, this author was not known for his egalitarian bent; he also wrote that the masses "must forever remain the masses. There would be no culture without kitchenmaids" (Treitschke 1963 [1916]: 24).
10 Admittedly, the liberal school of IR theory can be associated with a "transnationalist" literature which, itself, flows within the literature on new social movements, but we will leave most of this stream for the critical school because of the latter's explicit rejection of the state as a solution to crisis.
11 While Martin Wight (1991) extended the "Kantian school" to all revolutionary doctrines (including Marxism), Kantianism in IR theory is usually understood to refer to the main argument of Perpetual Peace (Kant 1795): that, at least in the short-term, war may be averted providing liberal principles are followed by the main actors in international politics. The right conditions included respect for the sanctity of the inviolable state; republicanism as an essential condition of peace; a peaceful world would by no means be united under a single jurisdiction; and peace would feed on the freedom of commerce and the general ideology of (rational) progress.
12 Realists, in contrast, tend to subscribe to a mercantilist vision of trade: it is intended to pursue the wealth and security of individual states.
13 The Western perspective is prevalent in Europe and in North America, and sees nature as alien and hostile, something that needs to be controlled; the Sineatic is from China, Korea, and Japan, and sees nature as beautiful and perfect but in need of transformation for us to love it. The Indian perspective – Hindu, Buddhist, and Jainist – sees nature as a mother/goddess.
14 The leading work in this area comes from a team of researchers led by Thomas Homer-Dixon (1991, 1994). We are not suggesting this work adopts an explicitly realist perspective, however; in fact, it emerges as critical of the shortcomings of that perspective.

2 Ecological thought: a synopsis

1 The concept of the "system" itself derives from the path-breaking work of Copernicus and Kepler on planetary motion. Kepler (along with Galileo) upheld the heliocentric view of the universe suggested by Copernicus, yet modifying some of its simplistic assumptions (see Kepler's

three laws) and, more importantly for our purposes, establishing clearly that the sun and its planets formed an interrelated system. Newton's mechanicism and materialism ultimately derive from the bases laid by these early giants.
2 As Aristotle wrote: "there is a limit, as there is to other things, plants, animals, implements; for none of these retain their natural power when they are too large or too small, but they either wholly lose their nature, or are spoiled" (in Coker 1914: 95).
3 As hinted above, Mill's father, James Mill, was a close associate of Bentham; John Stuart Mill was heavily influenced by both in his early years.
4 Tocqueville directly inspired Mill on this point.
5 For example, reading a good history book or taking a long walk in the woods. These cannot be equivalent to low-level pleasures, such as playing cards or using a new appliance. Since all individuals have experienced low-level pleasures, but only a minority have known higher-level pleasures, then, Mill argues, only the latter can testify to the difference – and their answer is inevitably to recognize the superiority of the more challenging activity.
6 Mill believed that all rational adults could effectively appreciate the "higher," or more "noble" things in life; education was the obvious means to that end, and so Mill was an ardent supporter of educational reform.
7 Bentham's pleasure principle, on the other hand (and as distinct from the classical liberal view of nature), remains basically unchallenged.
8 And so did not encompass distribution (Heilbroner 1961: 107–8).
9 On similar neo-Malthusian lines, the well-known work of Barbara Ward and René Dubos (1972) must be mentioned.
10 As a body, the state is also likely to collide with other states, and so a global Leviathan would seem recommended to palliate to that eventuality (although Hobbes refrained from this conclusion). We return to this theme in our discussion of realism in the next chapter.
11 Quoted in the *London Observer*, 14 April 1974.
12 She discusses this with reference to the anti-poaching program advocated by Northern environmentalists concerned with the survival of the African elephant. In the Kenyan case, the aircraft, radios, vehicles, night-goggles, and other anti-poaching equipment introduced by international wildlife conservation groups, such as WWF, IUCN, TRAFFIC, WCI, and the African Wildlife Federation probably contributed to the Kenyan government's oppression of various ethnic groups, including Somali pastoralists. She also looks at forest conservation policy in Indonesia in making the same point; see Peluso (1992) for an expanded treatment.
13 Guha (1990) is an exception.
14 Mumford's ecological creed is inherently anthropological: "Never was the ecological balance of nature, *and even more the integrity of cultures*, so violently upset during the last two centuries" (Mumford 1970: 379; emphasis added).

15 I.e. such technologies that are cheap, applicable in the small scale, and eliciting creativity on the part of the operator (Schumacher 1973: 34).
16 For instance, simple ("efficient," mass-market-oriented) monocultures will replace delicately balanced mixtures of crops. Conversely, highly complex power plants will be required to provide energy for modern societies' highly complex (and unending) sets of "needs."
17 Our discussion of Bookchin is inspired from his various works listed in our bibliography, all of which consistently present the same arguments and analyses that have brought Bookchin his popularity.

3 Realism and ecology

1 Most often, to be precise, realism is equated with the billiard-ball model, where states (billiard balls) bounce off one another. This is contrasted to the web-like model of complex interdependence or transnational relations. Both offer immediate visual recognition to students but are, of course, exceptionally simplistic. There is an inclination also, writes Michael Doyle, to identify realists by "their criticisms: the opponents of idealism or the critics of moralism, legalism, cosmopolitanism, or rationalism" (Doyle 1997: 42).
2 Doyle (1997) includes Jean-Jacques Rousseau as one of the founders of realism, and his account of Rousseau's "constitutional realism" is a fascinating read. However, we find Rousseau, one of the masters of paradox, difficult to place, since he has clearly contributed much to the liberal and, even, critical perspective as well.
3 Note that Thucydides can be read from a perspective that does not fixate on the balance-of-power politics of the ancient Greek city-state system, but on the social conventions and norms that dominated within it (see Garst 1989; and Elshtain 1997: 79–80).
4 It is understood that the term "anarchy" has a distinctive ring to realist scholars, describing the absence of a single authoritative government over states.
5 And so Treitschke (1963: 47) distances himself from Machiavelli: "It is not so much his total indifference to the means by which power is attained which repels us . . . but the fact that the power itself contains for him no deeper significance." Even Morgenthau (1946: 169), half-a-century later and in a striking (but rarely quoted) passage, stated that "the history of political thought is the history of the moral evaluation of power, and the scientism of Machiavelli-Hobbes is, in the history of humankind, merely an accident without consequences."
6 Wight could never be identified with either one of the three paradigms that he had popularized (realism, rationalism, and revolutionism).
7 Morgenthau and others opposed American involvement in Vietnam, for example.
8 Though the Machiavellian strain certainly equates the freedom, or political autonomy, and progress, or solidification of political rule, of leaders with the health of the state.
9 Morgenthau (1993: 37) writes: "The drives to live, to propagate and to dominate are common to all men." A classic response is in Mead (1973).

10 Treitschke (1963: 39–40) wrote: "War is the one remedy for an ailing nation . . . Heroism, bodily strength, and chivalrous spirit are essential to the character of a noble people . . . God above us will see to it that war shall return again."
11 Cf., on the one hand, the strikingly Durkheimian reference to controllable "social facts" (Morgenthau 1946: 128, 218–19), and, on the other, this fascinating (and neglected) passage: "Circumstances are infinite, are infinitely combined; are variable and transient; he who does not take them into consideration is not erroneous, but stark mad, . . . metaphysically mad" (Morgenthau 1946: 220).
12 Here Herz emerges as the American liberal as much as the intellectual realist: "atomic energy and other discoveries . . . have . . . opened up almost limitless possibilities. With the achievement of material abundance – . . . now apparently in the realm of the possible – a major obstacle in the way of the solution of the vicious circle of power and security competition would have disappeared" (Herz 1951: 251).
13 Cf. Herz (1951: 162): "Political realism shows that it is very unlikely that experts in any society would be allowed to retain at any length of time the amount of power which they request for the fulfilment of their aims."

4 Liberal IR theory and ecology

1 Zacher and Matthew conclude their exhaustive survey of contemporary liberal IR theory by insisting that all of the strands they identify "are ultimately about enhancing the security, prosperity, and human rights of individuals. While analyses often focus on state interests and interstate interactions, the lens through which they are evaluated by liberals is how they affect the material and moral conditions of people" (1995: 137).
2 However the evidence in terms of a neofunctionalist trend in the continuing evolution of the European Union, often considered the most fertile proving ground for what is admittedly a rather Eurocentric theoretical proposition, seems rather bleak (see in particular Huelshoff and Pfeiffer 1992).
3 "We shall offer no excuses for so frequently resolving questions of state policy into matters of pecuniary calculation. Nearly all the revolutions and great changes in the modern world have had a financial origin" (Cobden 1903: 238).
4 It is worth quoting Mazzini at length here (emphasis in original):

> Materialism has perpetuated our slavery by poisoning our souls with egotism and cowardice. Materialism . . . substituted for the idea that life is a mission and duty to be fulfilled, the idea that it is a *search after happiness*; . . . even this idea of happiness was corrupted into an idea of *pleasure*, of the happiness of a day or hour, to be bought by gold . . . Materialism broke asunder that social bond . . . to make the *individual* the center, end, and aim of our every endeavor, and substituted for the idea . . . of a providential educational

design and common progress, the cold lifeless conception of a *fatal* alternation of triumph and ruin, life and death.

(Mazzini 1945: 218)

5 Witness Mazzini's analysis of world events in the sixteenth and seventeenth centuries (emphasis in original): "the great problem of the day was a religious problem. . . . That which others called the *theory* of Machiavelli, appeared to me to be simply a history, the history of a period of corruption and degradation" (Mazzini 1945: 50).
6 Angell refuted criticisms by claiming never to have argued for the impossibility of war – only that war would be the path of the unenlightened, the emotional.
7 The teleology is strong, and revealing of the particular intolerance of this version of totalizing liberalism: "The programme of the world's peace . . . is our programme . . . the only possible programme"; and, further, "the culminating and final war for human liberty has come" (Link 1984: 536, 539).
8 David Long (1993) has pointed this out very effectively. Harold Laski (1967 [1925]: xx) would write that "given the class-relations of the modern state it is impossible to realise the ideal of an effective international community": rejecting here the transcendental character of law (as a mere expression of dominant economic interests), he rather believes in the institutionalized (and ongoing) process of "rational discussion" across borders, conditioning a "habit" of cooperation.
9 For an effective, succinct assessment of that research program, cf. Hansen (1969).
10 Mazrui's quest for a "world culture" would ensure that "Western culture . . . be infiltrated by non-Western values to help make the global pool of shared cultures less Eurocentric and more diversified" (Mazrui 1976: 11). Yet, practically, the objective will likely remain elusive, as Mazrui not only favors the sustainance of (Western-style) economic growth, but both reduces the concept of culture to selected languages (five) and ends up proposing, as a political structure, what amounts to a reformed UN (with all its bureaucracy and centralization). Kothari (1974) also directs some efforts toward a reformed world governmental structure, yet is much more critical of Western thought and much more clearly aware of the importance of diversity.
11 Note that even Northerners associated with WOMP or sympathetic to the WOMP agenda will often show the same partiality. Of course, the growth discourse will be qualified, but rarely do such authors specify when growth should stop, who should grow, and what should grow. See particularly Gurtov (1988: 172): "Economic growth that creates jobs and enhances life can proceed along with protection of the environment and conservation of resources." The language clearly preserves some of the basic commitments of the modern managerial society.
12 Kothari insists on agriculture as a "catalyst for growth" in the South and endorses the "green revolution" – which has since proven an ecological curse. The green revolution has also skewed landholding patterns:

Kothari recognizes this, but believes that the technology can be used rather harmlessly (Kothari 1974: 59).
13 Principles include: enforcement; the use of force only in self-defense; world judicial tribunals and organs of mediation/conciliation; a permanent world police force ("fully adequate"); complete disarmament of all nations; effective world machinery to bridge the world economic gap (cf. their World Development Authority); an active, virtually universal participation in the world authority; world law (on war prevention) to apply as well to individuals; restricting (for the moment) the powers of the world organization to issues of peace maintenance (Clark and Sohn 1958: xi–xiii).

5 Critical theory and ecology

1 This is usually dubbed the "third debate" in the field, following earlier debates between, on the one hand, idealism and realism, and, on the other, scientific and traditional approaches; see Lapid (1989). Critical IR theory as the attack against positivism is the main theme in Neufeld (1995).
2 Wendt's emphasis on "scientific realism" as an approach to causation is inspired by sources other than Giddens. The latter would not disagree with this basic idea, but would be more careful in upholding the power of science in social study (although he does declare himself a sociologist and not a philosopher).
3 Bergesen does not use the concept of structuration; yet his "globological" critique of Wallersteinian (1979) structuralism is reminiscent of Wendt, while he also specifically endorses a conception of process based on reciprocal influence between structure and agent (Bergesen 1990: 77).
4 Falk's "critical realists" are united by a common rejection of neorealism, and a sensitivity to alterations in historical conditions. He reflects the broader debate between so-called traditionalists and behaviorists that took place earlier in the IR theory discourse.
5 Note however that Stephen Gill argues Foucault's work emphasizes not structural change but "deep continuities in power relations and forms of subordination over time – that is after what he calls the 'epistemological break' of the Enlightenment and the onset of scientific rationality" (Gill, 1997: 8; see Foucault, 1972: 176–7).
6 See among others Sylvester (1994); Grant and Newland eds. (1991); Tickner (1992); Elshtain (1987); Peterson ed. (1992); Elshtain and Tobias eds. (1990); Enloe (1989); and the special 1992 edition (Vol. 21, No. 2) of *Millennium: Journal of International Studies*.
7 Dependency theory parallels, to some extent, Immanuel Wallerstein's treatment of the world capitalist system. It is not constructed, as such, as a critique of IR theory, with which it holds no dialogue. However, there is no doubt that International Relations can learn from the neoMarxian critique (cf. Kubalkova and Cruickshank 1985) and that the latter can be construed as critical theory (Leonard 1990).

8 Falk wrote:

> A world of sovereign states is unable to cope with endangered-planet problems ... such a system exhibits only a modest capacity for international cooperation and coordination. The distribution of power and authority, as well as the organization of human effort, is overwhelmingly guided by the selfish drives of nations.
> (Falk 1971: 37–38)

9 Shiva notes the crucial role played by the military-industrial complex in this regard. She gives the example of nitrogen, manufactured for explosives during the second world war and in sudden need for a market after the war, in this case, as a fertilizer; international agencies played a key role in subsidizing the product, giving it away in some cases (Shiva 1989: 69–70).
10 Structuration theory is not substantive. Wendt (1987: 355) states the point well: "[Structuration theory] does not tell us what particular kinds of agents or what particular kinds of structures to expect in any given concrete social system."
11 James Der Derian, in his spirited defense of postmodernism, writes of the need to shift, if not obliterate, "the positivist fact-value dichotomy in which the anguished social scientist seeks to expunge subjective factors from objective analysis" (Der Derian 1997: 61).

Bibliography

Adler, I. (1973) *Ecological Fantasies*, New York: Green Eagle.
Alger, C. and Mendlovitz, S. (1987) "Grass-roots Initiatives: The Challenges of Linkages", in S. Mendlovitz and R. B. J. Walker (eds) *Towards a Just World Peace: Perspectives from Social Movements*, London and Boston: Butterworths.
Alker, H. Jr. and Haas, P. (1993) "The Rise of Global Ecopolitics", in N. Choucri (ed.) *Global Accord: Environmental Challenges and International Responses*, Cambridge, Mass.: MIT Press.
Allaby, M. (1989) *Thinking Green: An Anthology of Essential Ecological Writings*, London: Barrie & Jenkins.
Amin, S. (1977) *Imperialism and Unequal Development*, New York: Monthly Review.
Angell, N. (1911) *The Great Illusion*, New York: Knickerbocker.
Aron, R. (1966) *Peace and War: A Theory of International Relations*, New York: Doubleday.
Ashley, R. (1989a) "Living on Border Lines: Man, Poststructuralism, and War", in J. Der Derian and M. Shapiro (eds) *International/Intertextual Relations: Postmodern Readings of World Politics*, Lexington: Lexington Books.
—— (1989b) "Imposing International Purpose: Notes on a Problematic of Governance" in J. Rosenau and E.- O. Czempiel (eds) (1992) *Governance Without Government: Order and Change in World Politics*, Cambridge: Cambridge University Press.
—— (1984) "The Poverty of Neo-Realism", *International Organization* 38, 225–86.
—— (1981) "Political Realism and Human Interests", *International Studies Quarterly* 25, 2: 204–36.
Atkinson, A. (1991) *Principles of Political Ecology*, London: Bellhaven.
Attfield, R. (1991) "Attitudes to Wildlife in the History of Ideas", *Environmental History Review* 15, 2: 71–8.
Baarschers, W. (1996) *Eco-facts and Eco-fiction: Understanding the Environmental Debate*, London: Routledge.

Bahro, R. (1972) *Socialism and Survival*, London: Heretic.
Bailey, R. (ed.) (1995) *The True State of the Planet*, New York: Free Press.
Barbier, E. (ed.) (1993) *Economics and Ecology: New Frontiers and Sustainable Development*, London: Chapman Hall.
Beckerman, W. (1993) *Small Is Stupid: Blowing the Whistle on the Greens*, London: Duckworth.
—— (1992) "Global Warming and International Action: An Economic Perspective", in A. Hurrell and B. Kingsbury (eds) *The International Politics of the Environment*, Oxford: Oxford University Press.
Beckmann, P. (1973) *Eco-Hysterics and the Technophobes*, Boulder: Golem.
Benedict, R. (1991) *Ozone Diplomacy: New Directions in Safeguarding the Planet*, Cambridge: Harvard University Press.
Bentham, J. (1952) *An Introduction to the Principles of Morals and Legislation*, New York: Harper & Brothers.
Benton, T. (1989) "Marxism and Natural Limits: An Ecological Critique and Reconstruction", *New Left Review* 178: 51–86.
Bergesen, A. (1990) "Turning World-System Theory on Its Head", *Theory, Culture and Society* 7, 2–3: 67–81.
Berman, M. (1981) *The Reenchantment of the World*, Ithaca: Cornell University Press.
Biehl, J. (1991) *Finding Our Way: Rethinking Ecofeminist Politics*, Montreal: Black Rose.
Biehl, J. and Staudenmaier, P. (1995) *Ecofascism: Lessons From the German Experience*, Edinburgh: AK Press.
Biersteker, T. (1989) "Critical Reflections on Post-Positivism in International Relations", *International Studies Quarterly* 33, 3: 263–7.
Blomstrom, M. and Hettne, B. (1984) *Development Theory in Transition: The Dependency Debate and Beyond*, London: Zed.
Boardman, R. (1997) "Environmental Discourse and International Relations Theory: Towards a Proto-theory of Ecosation", *Global Society* 11: 1, 31–44.
Bookchin, M. (1995) *The Philosophy of Social Ecology: Essays on Dialectical Naturalism*, Montreal: Black Rose.
—— (1992) *Urbanization Without Cities: The Rise and Decline of Citizenship*, Montreal: Black Rose.
—— (1991) *The Ecology of Freedom: The Emergence and Dissolution of Hierarchy*, Revised edition, Montreal: Black Rose.
—— (1991 [1980]) *Toward an Ecological Society*, Montreal: Black Rose.
—— (1990) "Recovering Evolution: A Reply to Eckersley and Fox", *Environmental Ethics*, 12, 3: 253–74.
—— (1989) *Remaking Society: Pathways to a Green Future*, Montreal: Black Rose.
—— (1986 [1973]) *The Limits of the City*, Montreal: Black Rose.
—— (1982) *The Ecology of Freedom*, Palo Alto: Cheshire.

—— (1971) *Post-Scarcity Anarchism*, Berkeley: Ramparts Press.
Bookchin, M. and Foreman, D. (1991) *Defending the Earth*, Montreal: Black Rose.
Botkin, D. (1990) *Discordant Harmonies: A New Ecology for the 21st Century*, New York: Oxford University Press.
Boulding, K. (1978) *Stable Peace*, Austin: University of Texas Press.
Bramwell, A. (1989) *Ecology in the 20th Century: A History*, New Haven: Yale University Press.
Brecher, J., Childs, J., and Cutler, J. (eds) (1993) *Global Visions: Beyond the New World Order*, Boston: South End.
Brenton, T. (1994) *The Greening of Machiavelli: The Evolution of International Environmental Politics*, London: Earthscan and Royal Institute of International Affairs.
Brown, L. *et al.* (1996) *State of the World 1996*, London: Earthscan.
Brown, S. (1995) *New Forces, Old Forces, and the Future of World Politics*, New York: HarperCollins.
Bull, H. (1977) *The Anarchical Society: A Study of Order in World Politics*, New York: Columbia University Press.
—— (1966) "International Theory: The Case for a Classical Approach", *World Politics* XVIII, 3: 361–77.
Bullard, R. (ed.) (1994) *Unequal Protection: Environmental Justice and Communities of Color*, San Francisco: Sierra Club.
Buzan, B. (1991) *People, States and Fear*, Second Ed., London: Harvester Wheatsheaf.
Buzan, B., Jones, C., and Little, R. (1993) *The Logic of Anarchy: Neorealism to Structural Realism*, New York: Columbia University Press.
Cahn, M. A. and O'Brien, R. (eds) (1996) *Thinking About the Environment: Readings on Politics, Property, and the Physical World*, Armonk: M.E. Sharpe.
Cahn, R. (ed.) (1985) *An Environmental Agenda for the Future*, Washington, DC: Island.
Caldwell, L. (1996) *International Environmental Policy: Emergence and Dimensions*, Durham, NC: Duke University Press.
—— (1990) *Between Two Worlds: Science, the Environmental Movement and Policy Choice*, Cambridge: Cambridge University Press.
—— (1988) "Beyond Environmental Diplomacy: The Changing Institutional Structure of International Cooperation", in J. Carroll (ed.), *International Environmental Diplomacy: The Management and Resolution of Transfrontier Environmental Problems*, Cambridge: Cambridge University Press.
Campanella, M. L. (1990) "Globalization: Processes and Interpretations", *World Futures* 30: 1–16.
Caporaso, J. (1997) "Across the Great Divide: Integrating Comparative and International Politics", *International Studies Quarterly* 41, 4: 563–92.
Capra, F. (1991) *The Tao of Physics*, Boston: Shambhala.

Cardoso, F. H. (1977) "The Consumption of Dependency Theory in the United States", *Latin American Research Review* 12, 3: 7–24.
Cardoso, F. H. and Faletto, E. (1979) *Dependency and Development in Latin America*, Berkeley: University of California Press.
Carpenter, G. (1997) "Redefining Scarcity: Marxism and Ecology Reconciled", *Democracy and Nature: The International Journal of Politics and Ecology* 3, 3: 129–53.
Carr, E. H. (1946) *The Twenty Years' Crisis, 1919–1939: An Introduction to the Study of International Relations*, London: Macmillan.
Carroll, J. (ed.) (1988) *International Environmental Diplomacy: The Management and Resolution of Transfrontier Environmental Problems*, Cambridge: Cambridge University Press.
Carson, R. (1962) *Silent Spring*, Boston: Houghton Mifflin.
Catton, W. (1980) *Overshoot: The Ecological Basis of Revolutionary Change*, Champaign, IL.: University of Illinois Press.
Ceadel, M. (1987) *Thinking about Peace and War*, Oxford: Oxford University Press.
Chase, A. (1989) "Missionaries of Environmentalism", *Orange County Register* August 6, 1989.
Chatterjee, P. and Finger, M. (1994) *The Earth Brokers: Power, Politics and World Development*, New York: Routledge.
Choucri, N. (ed.) (1993) *Global Accord: Environmental Challenges and International Responses*, Cambridge, Mass.: MIT Press.
Clark, G. and Sohn, L. (1958) *World Peace through World Law*, Cambridge, Mass: Harvard University Press.
Claude, I. (1962) *Power and International Relations*, New York: Random House.
Coates, K. (ed.) (1972) *Socialism and the Environment*, Nottingham: Spokesman.
Cobden, R. (1903) *Political Writings, Vol. 1*, London: T. Fisher Unwin.
Cohn, C. (1993) "Wars, Wimps and Women: Talking Gender and Thinking War", in M. Cooke and A. Wollacott (eds) *Gendering War Talk*, Princeton: Princeton University Press.
—— (1987) "Sex and Death in the Rational World of Defense Intellectuals", *Signs: Journal of Women in Culture and Society* 12, 4: 687–718.
Coker, F.W. (1914) *Readings in Political Philosophy*, New York: Macmillan.
Cole, H.S.D. *et al.* (1973) *Models of Doom: A Critique of "The Limits to Growth"*, London: Chatto and Windus.
Commission on Global Governance (1995) *Our Global Neighbourhood*, Oxford: Oxford University Press.
Commoner, B. (1976) *The Poverty of Power: Energy and the Economic Crisis*, New York: Alfred A. Knopf.
—— (1972) *The Closing Circle: Confronting the Environmental Crisis*, New York: Bantam.
Conca, K. (1995) "Greening the United Nations: Environmental Organizations and the UN System", *Third World Quarterly* 16, 3: 441–57.

Conca, K., Alberty, M., and Dabelko, G. (eds) (1995) *Green Planet Blues: Environmental Politics From Stockholm to Rio*, Boulder: Westview.

Cook, V. and Cook, E. (1988) "Romance and Resources", in P. Ehrlich and J. Holden (eds) *The Cassandra Conference: Resources and the Human Predicament*, College Station, Texas: A & M University Press.

Coward, H. (ed.) (1995) *Population, Consumption, and the Environment: Religious and Secular Responses*, New York: State University of New York Press.

Cox, R. (1994) "The Crisis in World Order and the Challenge to International Organization", *Cooperation and Conflict* 29: 99–113.

—— (1989) "Production, the State, and Change in World Order", in J. Rosenau and E.-O. Czempiel (eds) *Global Changes and Theoretical Challenges*, Lexington: Lexington Books.

—— (1987) *Production, Power and World Order: Social Forces in the Making of History*, New York: Columbia University Press.

—— (1983) "Gramsci, Hegemony and International Relations: An Essay in Method", *Millennium: Journal of International Studies* 12, 2: 162–75.

—— (1981) "Social Forces, States and World Order: Beyond International Relations Theory", *Millennium: Journal of International Studies* 10, 2: 126–55.

Cox, W. and Sjolander, C. T. (1994) "Critical Reflections on International Relations", in W. Cox and C. T. Sjolander (eds) *Beyond Positivism*, Boulder: Lynne Rienner.

DeBardeleben, J. and Hannigan, J. (eds) (1995) *Environmental Security and Quality After Communism: Eastern Europe and the Soviet Successor States*, Boulder: Westview.

Der Derian, J. (1997) "Post-Theory: The Eternal Return of Ethics in International Relations", in M. Doyle and G. J. Ikenberry (eds) *New Thinking in International Relations Theory*, Boulder: Westview.

Der Derian, J. and Shapiro, M. (eds) (1989) *International/Intertextual Relations: Post-Modern Readings of World Politics*, Toronto: Lexington.

Deudney, D. (1997) "Geopolitics and Change", in M. Doyle and G. J. Ikenberry (eds) *New Thinking in International Relations Theory*, Boulder: Westview.

—— (1992) "The Mirage of Eco-War: The Weak Relationship Among Global Environmental Change, National Security, and Interstate Violence", in I. Rowland and M. Greene (eds) *Global Environmental Change and International Relations*, London: Macmillan.

—— (1991) "Environment and Security: Muddled Thinking", *Bulletin of the Atomic Scientists* April: 22–8.

—— (1990) "The Case Against Linking Environmental Degradation and National Security", *Millennium: Journal of International Studies* 19, 3: 461–76.

Deutsch, K. *et al.* (1957) *Political Community and the North Atlantic Area*, Princeton: Princeton University Press.

Devall, B. and Sessions, G. (1985) *Deep Ecology: Living as if Nature Mattered*, Salt Lake City: Peregrine Smith.

Devetak, R. (1995) "The Project of Modernity and International Relations Theory", *Millennium: Journal of International Studies* 24, 1: 27–51.

Diamond, I. and Orenstein, G. (1990) *Reweaving the World: The Emergence of Ecofeminism*, San Francisco: Sierra Club.

Dobson, A. (1995) *Green Political Thought*, Second Edition, London: Routledge.

Dokken, K. and Graeger, N. (1995) *The Concept of Environmental Security*, Oslo: International Peace Research Institute.

Dowie, M. (1996) *Losing Ground: American Environmentalism at the Close of the Twentieth Century*, Cambridge: MIT Press.

Doyle, M. (1997) *Ways of War and Peace: Realism, Liberalism, and Socialism*, New York: W.W. Norton.

Doyle, M. and Ikenberry, G. J. (eds) (1997) *New Thinking in International Relations Theory*, Boulder: Westview.

Dryzek, J. (1987) *Rational Ecology: Environmental and Political Economy*, Oxford: Basil Blackwell.

Easlea, B. (1981) *Science and Sexual Oppression: Patriarchy's Confrontation With Women and Nature*, London: Weidenfeld & Nicholson.

Easterbrook, G. (1995) *A Moment on Earth: The Coming Age of Environmental Optimism*, New York: Viking.

Eckersley, R. (1992) *Environmentalism and Political Theory*, Albany: State University of New York Press.

—— (1988) "The Road to Ecotopia? Socialism Versus Environmentalism", *The Ecologist*, 18: 142–7.

Ehrenfeld, D. (1978) *The Arrogance of Humanism*, New York: Oxford University Press.

Ehrlich, P. (1970) *The Population Bomb*, New York: Ballantine.

Elshtain, J. (1997) "Feminist Inquiry and International Relations," in M. Doyle and G. J. Ikenberry (eds) *New Thinking in International Relations Theory*, Boulder: Westview.

—— (1987) *Women and War*, New York: Basic Books.

Elshtain, J. and Tobias, S. (eds) (1990) *Women, Militarism, and War*, Savage, MD: Rowman and Littlefield.

Enloe, C. (1989) *Bananas, Beaches, and Bases: Making Feminist Sense of International Politics*, London: Pandora.

Etzioni, A. (1965) *Political Unification: A Comparative Study of Leaders and Forces*, New York: Holt, Rinehart and Winston.

Evernden, L. (1993) *The Natural Alien: Humankind and Environment*, Toronto: University of Toronto Press.

Faber, D. (1993) *Environment Under Fire: Imperialism and the Ecological Crisis in Central America*, New York: Monthly Review.

Falk, R. (1997) "The Critical Realist Tradition and the Demystification of Power," in S. Gill and J. Mittelman (eds) *Innovation and Transformation in International Studies*, Cambridge: Cambridge University Press.
—— (1987) "The State System and Contemporary Social Movements", in S. H. Mendlovitz and R. B. J. Walker (eds) *Towards a Just World Peace: Perspectives from Social Movements*, London: Butterworths.
—— (1983) *The End of World Order*, New York: Holmes & Meier.
—— (1971) *This Endangered Planet: Prospects and Proposals for Human Survival*, New York: Vintage.
Falk, R. and Kim, S. (1982) *An Approach to World Order Studies and the World System*, New York: Institute for World Order.
Ferguson, Y. and Mansbach, R. (1991) "Between Celebration and Despair: Constructive Suggestions for Future International Theory", *International Studies Quarterly* 35, 4: 363–86.
Finger, M. (1991) "The Military, the Nation State and the Environment", *The Ecologist* 21, 5: 220–5.
Finger, M. and Kilcoyne, J. (1997) "Why Transnational Corporations are Organizing to 'Save the Global Environment'", *The Ecologist* 27, 4: 138–42.
Fisher, W. (ed.) (1995) *Toward Sustainable Development? Struggling Over India's Narmada River*, Armonk: M. E. Sharpe.
Foley, G. (1987) "Lewis Mumford–Philosopher of the Earth", *The Ecologist*, 17, 2: 109–15.
Foreman, D. (ed.) (1985) *Ecodefense: A Field Guide to Monkeywrenching*, Tucson: Ned Ludd.
Foucault, M. (1972) *The Archaeology of Knowledge*, trans. A. M. Sheridan Smith, London: Tavistock.
Fox, S. (1981) *John Muir and His Legacy: The American Conservation Movement*, New York: Little, Brown.
Fox, W. (1990) *Toward a Transpersonal Ecology: Developing New Foundations for Environmentalism*, Boston: Shambhala.
—— (1989) "The Deep Ecology-Ecofeminism Debate and Its Parallels", *Environmental Ethics*, 11, 1: 5–25.
Friberg, M. and Hettne, B. (1988) "Local Mobilization and World System Politics", *International Social Science Journal* 40, 117: 341–61.
Frome, M. (1996) *Chronicling the West: 30 Years of Environmental Writing*, Seattle: Mountaineers.
Gallie, W. B. (1978) *Philosophers of War and Peace*, Cambridge: Cambridge University Press.
Galtung, J. (1996) "On the Social Costs of Modernization: Social Disintegration, Atomie/Anomie and Social Development", *Development and Change* 27, 2: 379–413.
—— (1984) *There Are Alternatives! Four Roads to Peace and Security*, Nottingham: Spokesman.

—— (1971) "A Structural Theory of Imperialism", *Journal of Peace Research* 8, 2: 81–98.
—— (1964) "A Structural Theory of Aggression", *Journal of Peace Research* 1, 2: 95–119.
Gandhi, M. K. (1961) *Non-Violent Resistance*, New York: Schoken.
Gandy, M. (1996) "Crumbling Land: The Postmodernity Debate and the Analysis of Environmental Problems", *Progress in Human Geography* 20, 1: 23–40.
Gare, L. (1993) *Postmodernism and the Environmental Crisis*, London: Routledge.
Garst, D. (1989) "Thucydides and Neo-Realism", *International Studies Quarterly* 33, 1: 3–28.
George, J. (1994) *Discourses of Global Politics: A Critical (re) Introduction to International Relations*, Boulder: Lynne Rienner.
—— (1989) "International Relations and the Search for Thinking Space: Another View of the Third Debate", *International Studies Quarterly* 33, 3: 269–79.
George, J. and Campbell, D. (1990) "Patterns of Dissent and the Celebration of Difference: Critical Social Theory and International Relations", *International Studies Quarterly* 34, 3: 269–93.
Giddens, A. (1996) "Affluence, Poverty and the Idea of a Post-Scarcity Society", *Development and Change* 27, 2: 365–77.
—— (1991) "Structuration theory: past, present and future", in C. Bryant and D. Jary (eds) *Giddens' Theory of Structuration: A Critical Appraisal*, London: Routledge.
—— (1985) *A Contemporary Critique of Historical Materialism, Part II*, Berkeley: University of California Press.
—— (1984) *The Constitution of Society: Outline of the Theory of Structuration*, Berkeley: University of California Press.
Gilbert, F. (1971) 'Machiavelli: The Renaissance of the Art of War', in E. Mead Earle, ed., *Makers of Modern Strategy: Military Thought from Machiavelli to Hitler*, Princeton, Princeton University Press.
Gill, S. (1997) "Transformation and Innovation in the Study of World Order", in S. Gill and J. Mittelman (eds) *Innovation and Transformation in International Studies*, Cambridge: Cambridge University Press.
Gill, S. and Mittelman, J. (1997) *Innovation and Transformation in International Studies*, Cambridge: Cambridge University Press.
Gilpin, R. (1981) *War and Change in World Politics*, Cambridge: Cambridge University Press.
Goldsmith, E. (1988) "The Way: An Ecological World-View", *The Ecologist* 18, 4: 160–85.
Goldsmith, E. et al. (eds) (1995) *The Future of Progress: Reflections on Environment and Development*, Revised ed., Devon: Green.
Goldstein, J. (1988) *Long Cycles: Prosperity and War in the Modern Age*, New Haven: Yale University Press.

Gorz, A. (1980) *Ecology as Politics*, London: Pluto.
Grant, R. (1991) "The Sources of Gender Bias in International Relations Theory", in R. Grant and K. Newland (eds) *Gender and International Relations*, Bloomington: Indiana University Press.
Grant, R. and Newland, K. (eds) (1991) *Gender and International Relations*, Bloomington: Indiana University Press.
Gray, A. (1991) "The Impact of Biodiversity Conservation on Indigenous People" in V. Shiva *et al.* (eds) *Biodiversity: Social and Ecological Perspectives*, Penang: World Rainforest Movement.
Grieco, J. (1997) "Realist International Theory and the Study of World Politics", in M. Doyle and G. J. Ikenberry (eds) *New Thinking in International Relations Theory*, Boulder: Westview.
—— (1995) "Anarchy and the Limits of Cooperation: A Realist Critique of the Newest Liberal Institutionalism", in C. Kegley (ed.) *Controversies in International Relations Theory: Realism and the Neoliberal Challenge*, New York: St. Martin's.
Griffiths, M. (1992) *Realism, Idealism, and International Politics: A Reinterpretation*, London: Routledge.
Groom, A. and Taylor, P. (eds) (1975) *Functionalism: Theory and Practice in International Relations*, London: University of London Press.
Grubb, M. (1990a) "The Greenhouse Effect: Negotiating Targets", *International Affairs (London)* 66, 1: 67–89.
—— (1990b) *Energy Policy and the Greenhouse Effect*, London: Royal Institute for International Affairs.
Guha, R. (1990) "Toward a Cross-Cultural Environmental Ethic", *Alternatives* XV: 431–47.
—— (1989) *The Unquiet Woods: Ecological Change and Peasant Resistance in the Himalaya*, Delhi: Oxford University Press.
Gurtov, M. (1988) *Global Politics in the Human Interest*, Boulder: Lynne Rienner.
Haas, E. (1990) *When Knowledge is Power: Three Models of Change in International Organizations*, Berkeley: University of California Press.
—— (1983) "Words Can Hurt You: Or, Who said What to Whom About Regimes?", in S. Krasner (ed.) *International Regimes*, Ithaca: Cornell University Press.
—— (1975) *The Obsolescence of Regional Integration Theory*, Berkeley: Institute of International Studies, University of California.
—— (1964) *Beyond the Nation-State: Functionalism and International Organization*, Stanford: Stanford University Press.
—— (1958) *The Uniting of Europe: Political, Social and Economic Forces, 1950–1957*, Stanford: Stanford University Press.
Haas, P. (1990) *Saving the Mediterranean: The Politics of International Environmental Cooperation*, New York: Columbia University Press.

—— (1989) "Do Regimes Matter? Epistemic Communities and Mediterranean Pollution Control", *International Organization* 43, 3: 377–403.

Haas, P., Keohane, R., and Levy, M. (eds) (1993) *Institutions for the Earth: Sources of Effective International Environmental Protection*, Cambridge, Mass.: MIT Press.

Hair, J. (1991) "Nature Can Live with Free Trade", *New York Times* 19 May: Section 4, Page 17.

Halliday, F. (1994) *Rethinking International Relations*, Vancouver: University of British Columbia Press.

Hampson, F. (1988–89) "Climate change: building international coalitions of the like-minded", *International Journal*, XLV: 36–74.

Hansen, R. (1969) "Regional Integration: Reflections on a Decade of Theoretical Efforts", *World Politics*, 21: 242–71.

Hardin, G. (1974) "Living on a Lifeboat", *Bioscience* 24: 561–8.

—— (1973) *Exploring New Ethics for Survival: The Voyage of the Spaceship Beagle*, Baltimore: Penguin.

—— (1968) "The Tragedy of the Commons", *Science* 162: 1243–8.

Hartmann, F. (1957) *The Relations of Nations*, New York: Macmillan.

Hawkins, H. (1993) "Ecology", in R. Jackson (ed.) *Global Issues, 93–94*, Guilford, Conn.: Dushkin.

Hayward, T. (1994) *Ecological Thought: An Introduction*, Cambridge: Polity.

Hegedus, Z. (1989) "Social Movements and Social Change in Self-Creative Society: New Civil Initiatives in the International Arena", *International Sociology* 4, 1: 19–36.

Heilbroner, R. (1980) *An Inquiry Into the Human Prospect, Updated and Reconsidered for the 1980s*, New York: W.W. Norton.

—— (1961) *The Worldly Philosophers*, New York: Simon and Schuster.

Herz, J. (1951) *Political Realism and Political Idealism*, Chicago: University of Chicago Press.

Hessing, M. (1993) "Women and Sustainability: Ecofeminist Perspectives", *Alternatives: Perspectives on Society, Technology, and Environment* 19, 4: 14–21.

Hinsley, F. (1963) *Power and the Pursuit of Peace*, London: Cambridge University Press.

Hoffmann, M. (1987) "Critical Theory and the Inter-Paradigm Debate", *Millennium: Journal of International Studies* 16, 2: 231–49.

Hoffmann, S. (1965) *The State of War: Essays on the Theory and Practice of International Politics*, New York: Praeger.

Hollins, H., Powers, A., and Sommer, M. (1989) *The Conquest of War: Alternative Strategies for Global Security*, Boulder: Westview Press.

Holsti, K. J. (1985) *The Dividing Discipline: Hegemony and Diversity in International Theory*, Boston: Allen and Unwin.

Holsti, O. (1995) "Theories of International Relations and Foreign Policy: Realism and Its Challengers", in C. Kegley (ed.) *Controversies in Inter-*

national Relations Theory: Realism and the Neoliberal Challenge, New York: St. Martin's.
Homer-Dixon, T. (1994) "Environmental Scarcities and Violent Conflict: Evidence From Cases", *International Security* 19: 4–40.
—— (1991) "On the Threshold: Environmental Changes as Causes of Acute Conflict", *International Security* 16: 76–116.
—— (1990) *Environmental Change and Violent Conflict*, Cambridge: International Security Studies Program, American Academy of Arts and Sciences, Occasional Paper No. 4.
Huelshoff, M. and Pfeiffer, T. (1992) "Environmental Policy in the EC: Neo-Functionalist Sovereignty Transfer or Neo-Realist Gate-Keeping?", *International Journal* 47: 1: 136–58.
ICIDI (Independent Commission on International Development Issues) (1980) *North–South: A Programme for Survival*, London: Pan Books.
Isaac, R. and Isaac, E. (1985) *The Coercive Utopians*, Chicago: Regnery Gateway.
Jain, J. (ed.) (1996) *Environmental Stewardship and Sustainable Development*, New Delhi: Friedrich Ebert Stiftung.
Jary, D. (1991) "'Society as time-traveller': Giddens on historical change, historical materialism and the nation-state in world society", in C. Bryant and D. Jary (eds) *Giddens' Theory of Structuration: A Critical Appraisal*, London: Routledge.
Johansen, R. (1982) "Building a New International Security Order: Policy Guidelines and Recommendations", in C. Stephenson (ed.) *Alternative Methods for International Security*, Washington: University Press of America.
—— (1980) *The National Interest and the Human Interest: An Analysis of US Foreign Policy*, Princeton: Princeton University Press.
Jones, A. (1988) "From Fragmentation to Wholeness: A Green Approach to Science and Society, Part II", *The Ecologist*, 18, 1: 31–4.
—— (1987) "From Fragmentation to Wholeness: A Green Approach to Science and Society, Part I", *The Ecologist*, 17, 6: 237–40.
Kahler, M. (1997) "Inventing International Relations: International Relations Theory After 1945", in M. Doyle and G. J. Ikenberry (eds) *New Thinking in International Relations Theory*, Boulder: Westview.
Käkönen, J.(ed.) (1994) *Green Security or Militarized Environment?* Aldershot: Dartmouth.
Kamieniecki, S. (ed.) (1993) *Environmental Politics in the International Arena*, Albany: State University of New York Press.
Kant, I. (1983 [1795]) *Perpetual Peace and Other Essays*, trans. T. Humphrey, Indianapolis: Hackett.
Kaplan, M. (1966) "The New Great Debate: Traditionalism vs. Science in International Relations", *World Politics* XIX, 1: 1–20.
—— (1957) *System and Process in International Politics*, New York: Wiley.

Kaufman, W. (1994) *No Turning Back: Dismantling the Fantasies of Environmental Thinking*, New York: Basic Books.

Keeley, J. (1990a) "Toward a Foucauldian analysis of international regimes", *International Organization* 44, 1: 83–105.

—— (1990b) "The Latest Wave: A Critical Review of Regime Literature", in D. Haglund and M. Hawes (eds) *World Politics: Power, Interdependence and Dependence*, Toronto: Harcourt Brace Jovanovich.

Kegley, C. (ed.) (1995) *Controversies in International Relations Theory: Realism and the Neoliberal Challenge*, New York: St. Martin's.

Kegley, C. and Raymond, G. (1994) *A Multipolar Peace? Great Power Politics in the Twenty-first Century*, New York: St. Martin's.

Kelly, P. (1983) *Fighting for Hope*, trans. M. Howarth, Boston: South End.

Kennedy, P. (1987) *The Rise and Fall of the Great Powers: Economic Change and Military Conflict from 1500 to 2000*, New York: Random House.

Keohane R. (1989) *International Institutions and State Power*, Boulder: Westview.

—— (1984) *After Hegemony: Cooperation and Discord in the World Political Economy*, Princeton: Princeton University Press.

Keohane, R. and Nye, J. (1977) *Power and Interdependence*, Boston: Little, Brown.

—— (1971) *Transnational Relations and World Politics*, Cambridge, Mass: Harvard University Press.

Kober, S. (1990) "Idealpolitik", *Foreign Policy* 79: 3–24.

Kohr, L. (1957) *The Breakdown of Nations*, London: Routledge.

Kothari, R. (1974) *Footsteps into the Future*, New York: Free Press.

Krasner, S. (1985) *Structural Conflict: The Third World Against Global Liberalism*, Berkeley: University of California Press.

—— (1974) *Defending the National Interest: Raw Materials Investments and U.S. Foreign Policy*, Princeton: Princeton University Press.

Krasner, S. (ed.) (1983) *International Regimes*, Ithaca: Cornell University Press.

Kratochwil, F. (1989) *Rules, Norms, and Decisions: On the Conditions of Practical and Legal Reasoning in International Relations and Domestic Affairs*, Cambridge: Cambridge University Press.

Kropotkin, P. (1995) *Evolution and Environment*, in G. Woodcock (ed.), Montreal: Black Rose.

—— (1955 [1902]) *Mutual Aid: A Factor of Evolution*, Boston: Horizons.

Kubalkova, V. and Cruickshank, A. (1985) *Marxism and International Relations*, London: Routledge & Kegan Paul.

Kuehls, T. (1996) *Beyond Sovereign Territory: The Space of Ecopolitics*, Minneapolis: University of Minnesota Press.

Laferrière, E. (1996) "Emancipating International Relations Theory: An Ecological Perspective", *Millennium: Journal of International Studies* 25, 1: 53–75.

—— (1995) *The Failure of Peace: An Ecological Critique of International Relations Theory*, Ph.D. dissertation, McGill University.
—— (1994) "Environmentalism and the Global Divide", *Environmental Politics* 3, 1: 91–113.
Lamb, R. (1996) *Promising the Earth*, London: Routledge, 1996.
Lapid, Y. (1989) "The Third Debate: On the Prospects of International Theory in a Post-Positivist Era", *International Studies Quarterly* 33, 3: 235–54.
Laski, H. (1967 [1925]) *A Grammar of Politics*, London: Allen and Unwin.
Leff, E. (1993) "Marxism and the Environmental Question", *Capitalism, Nature, Socialism* 4, 1.
Leiss, W. (1972) *The Domination of Nature*, New York: G. Braziller.
Leith, J., Price, R., and Spencer, J. (eds) (1995) *Planet Earth: Problems and Prospects*, Montreal and Kingston: McGill-Queen's University Press.
Lélé, S. (1991) 'Sustainable Development: A Critical Review', *World Development*, 19: 6, 607–21.
Leonard, S. (1990) *Critical Theory in Political Practice*, Princeton: Princeton University Press.
Leopold, A. (1966 [1949]) *A Sand County Almanac*, New York: Ballantine.
Levy, M., Keohane, R., and Haas, P. (eds) (1993) *Institutions for the Earth: Sources of Effective Environmental Protection*, Cambridge, Mass.: MIT Press.
Link, A. (ed.) (1984) *The Papers of Woodrow Wilson, Vol. 45*, Princeton: Princeton University Press.
Linklater, A. (1990) *Beyond Realism and Marxism: Critical Theory and International Relations*, London: Macmillan.
—— (1990 [1982]) *Men and Citizens in the Theory of International Relations*, London: Macmillan.
Lipietz, A. (1995) *Green Hopes: The Future of Political Ecology*, trans. M. Slater, Cambridge: Polity.
Lipschutz, R. (1996) *Global Civil Society and Global Enviromental Governance: The Politics of Nature From Place to Planet*, Albany: State University of New York Press.
Lipschutz, R. and Conca, K. (eds) (1993) *The State and Social Power in Global Environmental Politics*, New York: Columbia University Press.
Litfin, K. (1997) "Sovereignty in World Ecopolitics", *Mershon International Studies Review* 41, 2: 167–204.
—— (1994) *Ozone Discourses: Science and Politics in Global Environmental Cooperation*, New York: Columbia University Press.
—— (1993) "Ecoregimes: Playing Tug of War with the Nation-State", in R. Lipschutz and K. Conca (eds) *The State and Social Power in Global Environmental Politics*, New York: Columbia University Press.
Little, R. (1991) "Liberal hegemony and the realist assault: Competing ideological theories of the state", in M. Banks and M. Shaw (eds) *State and Society in International Relations*, Hemel Hempstead: Harvester Wheatsheaf.

Long, D. (1993) "International Functionalism and the Politics of Forgetting", *International Journal* XLVIII: 356–79.
Lovelock, J. (1979) *Gaia: A New Look at Life on Earth*, Oxford: Oxford University Press.
Luke, T. (1996) "Liberal Society and Cyborg Subjectivity: The Politics of Environments, Bodies, and Nature", *Alternatives* 21, 1: 1–30.
McAllister, P. (ed.) (1982) *Reweaving the Web of Life: Feminism and Nonviolence*, Philadelphia: New Society Publishers.
McCormick, J. (1989) *Reclaiming Paradise: The Global Environmental Movement*, Bloomington: Indiana University Press.
McKinlay, R. and Little, R. (1986) *Global Problems and World Order*, London: Frances Pinter.
MacNeill, J. (1989–90) "The Greening of International Relations", *International Journal* XLV: 1–35.
Mahony, R. (1992) "Debt-for-Nature Swaps: Who Really Benefits?", *The Ecologist* 22, 3: 97–103.
Manes, C. (1990) *Green Rage: Radical Environmentalism and the Unmaking of Civilization*, Boston: Little, Brown.
Mann, D. (1991) "Environmental Learning in a Decentralized Political World", *Journal of International Affairs* 44, 2: 301–38.
Manning, D. (1976) *Liberalism*, London: J.M. Dent.
Mansbach, R. and Vasquez, J. (1981) *In Search of Theory: A New Paradigm for Global Politics*, New York: Columbia University Press.
Martell, L. (1994) *Ecology and Society: An Introduction*, Cambridge: Polity.
Marx, K. (1959) *Capital, Volume III*, Moscow: Foreign Publishing House.
Mathews, J. T. (1989) "Redefining Security", *Foreign Affairs* 68, 2: 162–77.
Mazrui, A. (1976) *A World Federation of Cultures: An African Perspective*, New York: Free Press.
Mazzini, G. (1945) *Selected Writings*, ed. N. Gangulee, London: Lindsay Drummond.
Mead, M. (1973) "Warfare Is Only an Invention, Not a Biological Necessity", reprinted in C. Beitz and T. Herman (eds) *Peace and War*, San Francisco: W. H. Freeman.
Meadows, D. et al. (1972) *The Limits to Growth*, New York: Potomac Associates.
Mearsheimer, J. (1990) "Back to the Future: Instability in Europe after the Cold War", *International Security* 15: 5–56.
Mellos, K. (1988) *Perspectives on Ecology*, London: Macmillan.
Merchant, C. (1996) *Earthcare: Women and the Environment*, New York: Routledge.
—— (1989) *Ecological Revolutions: Nature, Gender, and Science in New England*, Chapel Hill: University of North Carolina Press.
—— (1980) *The Death of Nature: Women, Ecology, and the Scientific Revolution*, New York: Harper and Row.

Middleton, N., O'Keefe, P., and Mayo, S. (1993) *Tears of the Crocodile: From Rio to Reality in the Developing World*, London: Pluto.
Mies, M. (1993) *Patriarchy and Accumulation on a World Scale: Women in the International Division of Labour*, London: Zed.
Mies, M. and Shiva, V. (1993) *Ecofeminism*, London: Zed.
Mikesell, R. (1992) *Economic Development and the Environment: A Comparison of Sustainable Development with Conventional Development Economics*, London: Mansell.
Milbrath, L. (1984) *Environmentalists: Vanguard for a New Society*, Albany: State University of New York Press.
Milner, H. (1992) "International Theories of Cooperation Among Nations: Strengths and Weaknesses", *World Politics* 44, 3: 466–96.
Mische, G. and Mische, P. (1977) *Toward a Human World Order*, New York: Paulist Press.
Mische, P. (1989) "Ecological Security and the Need to Reconceptualize Sovereignty", *Alternatives* 14: 389–427.
—— (1982) "Revisioning National Security: Toward a Viable World Security System", in C. Stephenson (ed.) *Alternative Methods for International Security*, Washington: University Press of America.
Mitchell, R. (1994) *Intentional Oil Pollution at Sea: Environmental Policy and Treaty Compliance*, Cambridge, Mass.: MIT Press.
Mitrany, D. (1966) *A Working Peace System*, Chicago: Quadrangle.
Morgenthau, H. (1993) *Politics Among Nations* [brief. ed.], New York: McGraw-Hill.
—— (1985) *Politics Among Nations: The Struggle for Power and Peace*, Sixth Ed., New York: Knopf.
—— (1948) *Politics Among Nations: The Struggle for Power and Peace*, New York: Knopf.
—— (1946) *Scientific Man Versus Power Politics*, Chicago: University of Chicago Press.
Morris, D. (1990) "Free Trade: The Great Destroyer", *The Ecologist* 20, 5: 190–5.
Morrison, R. (1995) *Ecological Democracy*, Boston: South End.
Mumford, L. (1970) *The Myth of the Machine: The Pentagon of Power*, New York: Harcourt Brace Jovanovich.
—— (1964) "Authoritarian and Democratic Technics", *Technology and Culture* 5: 1–8.
Nadelmann, E. (1990) "Global Prohibition Regimes: The Evolution of Norms in International Relations", *International Organization*, 44, 4: 479–526.
Naess, A. (1989) *Ecology, Community and Lifestyle*, trans. and ed. D. Rothenberg, Cambridge: Cambridge University Press.
—— (1972) "The Shallow and the Deep, Long-range Ecology Movement: A Summary", *Inquiry*, 16: 95–100.

Nagel, S. (1991) *Global Policy Studies: International Interaction Toward Improving Public Policy*, New York: St. Martin's.

Nash, R. (1989) *The Rights of Nature: A History of Environmental Ethics*, Madison: University of Wisconsin Press.

Neufeld, M. (1995) *The Restructuring of International Relations Theory*, Cambridge: Cambridge University Press.

Neuhaus, R. (1971) *In Defense of People: Ecology and the Seduction of Radicalism*, New York: Macmillan.

Niebuhr, R. (1960 [1932]) *Moral Man and Immoral Society*, New York: Charles Scribner's Sons.

—— (1945) *The Children of Light and the Children of Darkness*, London: Nisbet and Co.

Norton, B. (1991) *Toward Unity Among Environmentalists*, New York: Oxford University Press.

Nye, J. (1990) *Bound to Lead: The Changing Nature of American Power*, New York: Basic.

Olsen, M. (1971) "Increasing the Incentives for International Cooperation", *International Organization* 25: 866–74.

Onuf, N. and Johnson, T. (1995) "Peace in the Liberal World: Does Democracy Matter?", in C. Kegley (ed.) *Controversies in International Relations Theory: Realism and the Neoliberal Challenge*, New York: St. Martin's.

Ophuls, W. (1977) *Ecology and the Politics of Scarcity Revisited: The Unraveling of the American Dream*, New York: W. H. Freeman.

—— (1992) *Ecology and the Politics of Scarcity: A Prologue to a Political Theory of the Steady State*, San Francisco: W.H. Freeman.

O'Riordan, T. (1981) *Environmentalism*, Second Ed., London: Pion.

Orr, D. and Hill, S. (1979) "Leviathan, the Open Society, and the Crisis of Ecology", in D. Orr and M. Soroos (eds) *The Global Predicament*, Chapel Hill: University of North Carolina Press.

Ostrom, E. (1991) *Governing the Commons: The Evolution of Institutions for Collective Action*, Cambridge: Cambridge University Press.

Oye, K. (ed.) (1986) *Cooperation Under Anarchy*, Princeton: Princeton University Press.

Paehlke, R. (1989) *Environmentalism and the Future of Progressive Politics*, New Haven: Yale University Press.

Page, D. (1989) "Debt-for-Nature Swaps: Experience Gained, Lessons Learned", *International Environmental Affairs* 1, 4: 275–88.

Paterson, M. (1996) "IR Theory: Neorealism, Neoinstitutionalism and the Climate Change Convention", in J. Vogler and M. Imber (eds) *The Environment and International Relations*, London: Routledge.

Peluso, N. (1993) "Coercing Conservation: The Politics of State Resource Control", in R. Lipschutz and K. Conca (eds) (1993) *The State and Social Power in Global Environmental Politics*, New York: Columbia University Press.

—— (1992) *Rich Forests, Poor People: Resource Control and Resistance in Java*, Berkeley: University of California Press, 1992.
Pepper, D. (1993) *Eco-Socialism: From Deep Ecology to Social Justice*, London: Routledge.
—— (1989) *The Roots of Modern Environmentalism*, London: Routledge.
Peterson, V. S. (1997) "Whose Crisis? Early and Post-modern Masculinism," in S. Gill and J. Mittelman (eds) *Innovation and Transformation in International Studies*, Cambridge: Cambridge University Press.
—— (1992) "Transgressing Boundaries: Theories of Knowledge, Gender and International Relations", *Millennium: Journal of International Studies* 21, 2: 183–206.
Peterson, V. S. (ed.) (1992) *Gendered States: Feminist (Re) Visions of International Relations Theory*, Boulder: Lynne Rienner.
Pirages, D. (1991) "Environmental Security and Social Evolution", *International Studies Notes* 16, 1: 8–12.
—— (1983) "The Ecological Perspective and the Social Sciences", *International Studies Quarterly* 27, 3: 243–55.
—— (1978) *The New Context for International Relations: Global Ecopolitics*, North Scituate, Mass.: Duxbury Press.
Plumwood, V. (1993) *Feminism and the Mastery of Nature*, London: Routledge.
—— (1992) "Feminism and the Ecofeminism: Beyond the Dualistic Assumptions of Women, Men and Nature", *The Ecologist*, 22, 1: 8–13.
—— (1986) "Ecofeminism: An Overview and Discussion of Positions and Arguments", *Australian Journal of Philosophy* 64: 120–38.
Ponting, C. (1991) *A Green History of the World*, New York: Penguin.
Porritt, J. (1984) *Seeing Green: The Politics of Ecology Explained*, New York: Basil Blackwell.
Porter, G. and Brown, J. W. (1991) *Global Environmental Politics*, Boulder: Westview.
Ravindra, R. (1991) *Science and Spirit*, New York: Paragon House.
Ray, D. L. and Guzzo, L. (1993) *Environmental Overkill*, New York: HarperCollins.
Redclift, M. (1987) *Sustainable Development: Exploring the Contradictions*, New York: Methuen.
Richardson, L. (1960) *Arms and Insecurity: A Mathematical Study of the Causes and Origins of War*, Pittsburgh: Boxwood Press.
Ridgeway, J. (1970) *The Politics of Ecology*, New York: E. P. Dutton.
Riggs, R. and Plano, J. (1988) *The United Nations: International Organizations and World Politics*, Chicago: Dorsey.
Ritchie, M. (1992) "Free Trade vs Sustainable Agriculture: The Implications of NAFTA", *The Ecologist* 22, 5: 221–7.
Rittberger, V. and Mayer, P. (eds) (1993) *Regime Theory and International Relations*, Oxford: Clarendon.

Robertson, R. (1990) "Mapping the Global Condition: Globalization as the Central Concept", *Theory, Culture and Society* 7, 2–3: 15–30.
Roosevelt, G. (1990) *Reading Rousseau in the Nuclear Age*, Philadelphia: Temple University Press.
Rosenau, J. (1997) 'The Complexities and Contradictions of Globalization', *Current History* 96: 613, 360–64.
—— (1990) *Turbulence in World Politics: A Theory of Change and Continuity*, Princeton: Princeton University Press.
—— (1986) "Before Cooperation: Hegemons, Regimes and Habit-Driven Actors in World Politics", *International Organization* 40: 849–94.
—— (1984) "A Pre-Theory Revisited: World Politics in an Era of Cascading Interdependence", *International Studies Quarterly* 28, 3: 245–305.
—— (1981) *The Study of World Interdependence*, New York: Nichols.
—— (1980) *The Study of Global Interdependence*, London: Pinter.
—— (1969) *Linkage Politics: Essays on the Convergence of National and International Systems*, New York: Free Press.
Rosenau, J. and Czempiel, E.- O. (eds) (1992) *Governance Without Government: Order and Change in World Politics*, Cambridge: Cambridge University Press.
Rosenau, P. (1990) "Once Again Into the Fray: International Relations Confronts the Humanities", *Millennium: Journal of International Studies* 19, 1: 83–110.
Roszak, T. (1992) *The Voice of the Earth*, New York: Simon & Schuster.
—— (1972) *Where the Wasteland Ends: Politics and Transcendence in Postindustrial Society*, Garden City: Doubleday.
Rowland, I. and Greene, M. (eds) (1992) *Global Environmental Change and International Relations*, London: Macmillan.
Ruggie, J. G. (1989) "International Structure and International Transformation: Space, Time, and Method", in J. Rosenau and E.-O. Czempiel (eds) *Global Changes and Theoretical Challenges*, Lexington: Lexington Books.
Russett, B. (1982) "Causes of Peace", in C. Stephenson (ed.) *Alternative Methods for International Security*, Washington: University Press of America.
Sachs, W. (ed.) (1993) *Global Ecology: A New Arena of Political Conflict*, London: Zed.
—— (1992) "Environment", in same (ed.) *The Development Dictionary*, London: Zed.
Sagoff, M. (1995) "Can Environmentalists Be Liberal?", in R. Elliot (ed.) *Environmental Ethics*, New York: Oxford University Press.
Sale, K. (1996) "Natural Neighbors", *The New Internationalist* 278: 17.
—— (1985) *Dwellers in the Land: The Bioregional Vision*, Philadelphia: New Society.
—— (1974) "Mother of All: An Introduction to Bioregionalism", in S. Kumar (ed.) *The Schumacher Lectures: Volume II*, London: Blond and Briggs.

Sand, P. (1990) *Lessons Learned in Global Environmental Governance*, Washington: World Resources Institute.
Sarkar, S. (1993) *Green Alternative Politics in West Germany*, Two Volumes, Tokyo: United Nations University Press.
Saurin, J. (1996) "International Relations, Social Ecology and the Globalisation of Environmental Change", in R. Vogler and M. Imber (eds) *The Environment and International Relations*, London: Routledge.
Scherer, D., and Attig, T. (eds) (1983) *Ethics and the Environment*, Englewood Cliffs: Prentice-Hall.
Schmidt, A. (1971) *The Concept of Nature in Marx*, London: NLB.
Schulze, P. (ed.) (1996) *Engineering Within Ecological Constraints*, Washington: National Academy Press.
Schüking, H. and Anderson, P. (1991) "Voices Unheard and Unheeded", in V. Shiva *et al.* (eds) *Biodiversity: Social and Ecological Perspectives*, Penang: World Rainforest Movement.
Schumacher, E. F. (1973) *Small is Beautiful: Economics as if People Mattered*, New York: Harper and Row.
Scott, A. (1982) *The Dynamics of Interdependence*, Chapel Hill: University of North Carolina Press.
Sessions, G. (ed.) (1995) *Deep Ecology for the 21st Century*, Boston: Shambhala.
Sharp, G. (1973) *The Politics of Nonviolent Action*, Boston: P. Sargent.
Shields, L. and Ott, M. (1974) "The Environmental Crisis: International and Supranational Approaches", *International Relations* 4, 6: 627–48.
Shiva, V. (1993) *Monocultures of the Mind*, London: Zed.
—— (1992) "Global Bullies", *New Internationalist*, 230: 26.
—— (1991) "Biodiversity, Biotechnology and Profits", in V. Shiva *et al. Biodiversity: Social and Ecological Perspectives*, Penang: World Rainforest Movement.
—— (1989) *The Violence of the Green Revolution: Ecological Degradation and Political Conflict in Punjab*, Dehra Dun: Research Foundation for Science and Technology.
—— (1988) *Staying Alive: Women, Ecology, and Development in India*, New Delhi: Kali for Women.
Simmons, I. G. (1993) *Interpreting Nature: Cultural Constructions of the Environment*, London: Routledge.
Simon, J. (1996) *The Ultimate Resource Two*, Princeton: Princeton University Press.
Simon, J. and Kahn, H. (eds) (1984) *The Resourceful Earth: A Response to Global 2000*, Oxford: Blackwell.
Singer, P. (1990 [1975]) *Animal Liberation*, New York: Avon.
Smith, N. (1990) *Uneven Development: Nature, Capital and the Production of Space*, Oxford: Blackwell.
Smoke, R. and Harman, W. (1987) *Paths to Peace: Exploring the Feasibility of Sustainable Peace*, Boulder: Westview.

Snyder, G. (1990) *The Practice of the Wild*, Berkeley: North Point.
Soroos, M. (1986) *Beyond Sovereignty: The Challenge of Global Policy*, Columbia: University of South Carolina Press.
—— (1985) "Environmental Values", "Environmental Policies", and "The Future of the Environment", in K. Dahlberg (ed.) *Environment and the Global Arena*, Durham: Duke University Press.
Spector, B., Sjostedt, G., and Zartman, W. (eds) (1994) *Negotiating International Regimes: Lessons Learned From the United Nations Conference on Environment and Development*, London: Graham & Trotman.
Speth, G. (1990) "Toward a North–South Compact on the Environment", *Environment* 32, 5: 16–20, 40–3.
Sprout, H. and Sprout, M. (1978) *The Context of Environmental Politics: Unfinished Business for America's Third Century*, Lexington: University Press of Kentucky.
—— (1971) *Toward a Politics of the Planet Earth*, New York: Van Nostrand.
—— (1965) *The Ecological Perspective on Human Affairs, with Special Reference to International Politics*, Princeton: Princeton University Press.
Stoett, P. (1998 forthcoming) "Ecocide Revisited: Two Understandings", *Environment and Security*.
—— (1994) "The Environmental Enlightenment: Security Analysis Meets Ecology", *Coexistence: A Review of East–West and Development Issues* 31: 127–46.
Strange, S. (ed.) (1984) *Paths to International Political Economy*, London: George Allen and Unwin.
Sylvan, R. and Bennett, D. (1988) "Taoism and Deep Ecology", *The Ecologist*, 18, 4/5: 148–59.
Sylvester, C. (1994) *Feminist Theory and International Relations in a Postmodern Era*, Cambridge: Cambridge University Press.
Taylor, B. (ed.) (1993) *Ecological Resistance Movements: The Global Emergence of Radical and Popular Environmentalism*, Albany: State University of New York Press.
—— (1991) "The Religion and Politics of Earth First!", *The Ecologist*, 21, 6: 258–66.
Taylor, L. (1996) "Sustainable Development: An Introduction", *World Development* 24, 2: 215–25.
Thiele, L. (1993) "Making Democracy Safe for the World: Social Movements and Global Politics", *Alternatives* 18: 273–305.
Thomas, W. (1995) *Scorched Earth: The Military's Assault on the Environment*, Gabriola Island: New Society.
Thoreau, H. D. (1962) *Walden and Other Writings*, ed. J. Wood Krutch, New York: Bantam.
Thucydides (1972) *The Peloponnesian War*, trans. by Rex Warner, Harmondsworth: Penguin.
Tickner, J. (1992) *Gender in International Relations: Feminist Perspectives on Achieving Global Security*, New York: Columbia University Press.

—— (1991) "Hans Morgenthau's Principles of Political Realism: A Feminist Reformulation", in R. Grant and K. Newland (eds) *Gender and International Relations*, Bloomington: Indiana University Press.
Tobias, M. (1994) *World War III: Population and the Biosphere at the End of the Millennium*, Sante Fe: Bear and Company.
Treitschke, H. (1963 [1916]) *Politics*, New York: Harcourt, Brace and World.
Vajk, P. (1978) *Doomsday Has Been Cancelled*, Culver City: Peace Press.
Vernadsky, V. I. (1945) "The Biosphere and the Noosphere", *American Scientist*, 33: 4.
Vernon, R. (1971) *Sovereignty at Bay: The Multinational Spread of US Enterprises*, New York: Basic.
Viotti, P. and Kauppi, M. (1987) *International Relations Theory: Realism, Pluralism, Globalism*, New York: Macmillan.
Walker, R. B. J. (1993) *Inside/Outside: International Relations as Political Theory*, Cambridge: Cambridge University Press.
—— (1990) "Security, Sovereignty, and the Challenge of World Politics", *Alternatives*, XV: 3–27.
Wallerstein, I. (1979) *The Capitalist World Economy*, Cambridge: Cambridge University Press.
Waltz, K. (1979) *Theory of International Politics*, Reading, Mass.: Addison-Wesley.
—— (1959) *Man, the State and War: A Theoretical Analysis*, New York: Columbia University Press.
Wapner, P. (1997) "Environmental Ethics and Global Governance: Engaging the International Liberal Tradition", *Global Governance* 3: 213–31.
—— (1996) *Environmental Activism and World Civic Politics*, Albany: State University of New York Press.
Ward, B. and Dubos, R. (1972) *Only One Earth: The Care and Maintenance of a Small Planet*, New York: W. W. Norton.
WCED [World Commission on Environment and Development] (1987) *Our Common Future*, Oxford: Oxford University Press.
Weber, S. (1997) "Institutions and Change", in M. Doyle and G. J. Ikenberry (eds) *New Thinking in International Relations Theory*, Boulder: Westview.
Weisberg, B. (1971) *Beyond Repair: The Ecology of Capitalism*, Boston: Beacon.
Weiss, E. B. (1993) "Intergenerational Equity: Toward an International Legal Framework", in N. Choucri (ed.) *Global Accord: Environmental Challenges and International Responses*, Cambridge: MIT Press.
—— (1990) "Our Rights and Obligations to Future Generations for the Environment", *American Journal of International Law* 84, 1: 190–212.
—— (1989) *In Fairness to Future Generations: International Law, Common Patrimony, and Intergenerational Equity*, Dobbs Ferry, New York: Transnational and United Nations University.

Wendt, A. (1992) "Anarchy Is What States Make of It: The Social Construction of Power Politics", *International Organization* 46: 391–426.
—— (1987) "The Agent-Structure Problem in International Relations Theory", *International Organization* 41, 3: 335–70.
Wendt, A. and Duvall, R. (1989) "Institutions and International Order", in J. Rosenau and E.-O. Czempiel, *Global Changes and Theoretical Challenges*, Lexington: Lexington Books.
Westing, A. (1986) *Global Resources and International Conflict: Environmental Factors in Strategic Policy and Action*, New York: Oxford University Press.
Westing, A. (ed.) (1988) *Cultural Norms, War and the Environment*, Oxford: Oxford University Press.
Weston, B. (ed.) (1990) *Alternative Security: Living Without Nuclear Deterrence*, Boulder: Westview.
White, L. (1967) "The Historical Roots of Our Ecologic Crisis", *Science*, 155: 1203–7.
Wight, M. (1991) *International Theory: The Three Traditions*, Leicester: Leicester University Press.
Wijkman, P. M. (1982) "Managing the Global Commons", *International Organization* 36, 3: 511–36.
Winner, L. (1986) *The Whale and the Reactor: A Search for Limits in an Age of High Technology*, Chicago: University of Chicago Press.
Wolfers, A. (1962) *Discord and Collaboration: Essays on International Politics*, Baltimore: Johns Hopkins Press.
Woodcock, G. (1992) *Anarchism and Anarchists: Essays*, Kingston: Quarry.
Woodward, B. (1977) "Institutionalization of Nonviolence", *Alternatives* 3: 49–73.
Wooley, W. (1988) *Alternatives to Anarchy: American Supranationalism Since World War II*, Bloomington: Indiana University Press.
Worster, D. (1985) *Nature's Economy: A History of Ecological Ideas*, Cambridge: Cambridge University Press.
Young, J. (1990) *Post-Environmentalism*, London: Belhaven.
Young, O. (1994) *International Governance: Protecting the Environment in a Stateless Society*, Ithaca: Cornell University Press.
—— (1989a) *International Cooperation: Building Regimes for Natural Resources and the Environment*, Ithaca: Cornell University Press.
—— (1989b) "The Politics of Regime Formation: Managing Natural Resources and the Environment", *International Organization*, 43, 3: 349–75.
Zacher, M. and Matthew, R. (1995) "Liberal International Theory: Common Threads, Divergent Strands", in C. Kegley (ed.) *Controversies in International Relations Theory: Realism and the Neoliberal Challenge*, New York: St. Martin's.
Zimmerman, M. (1994) *Contesting Earth's Future: Radical Ecology and Postmodernity*, Berkeley: University of California Press.

Index

acidification 3
Agenda 21 2, 175n1
Amin, S. 143
anarchism 63–5
anarchy, in IR 8
Ancient Egypt 57
Ancient Greece 22
Angell, N. 112, 118–19, 126–7, 135
Anthropocentrism 68
Aristotle 13, 21, 23
Aron, R. 8, 81, 88, 91, 93
Arthashastra 7
Ashley, R. 146–7, 148, 155
Audubon Society 39
autarky 94

Bahro, R. 47
Bailey, R. 31
Bakunin, M. 55, 69
balance of power 8, 116
Basel Convention 40, 41
behavioralism 6, 80, 81, 121
Bentham, J. 27, 30, 32, 65
Biehl, J. 47
biodiversity loss 3, 41
bioregionalism 53, 63, 66
biospheric egalitarianism 62
birth control programmes 34
Bismarck, O. 78
Boardman, R. 19, 155–6
Bookchin, M. 16, 52, 55, 56, 59, 136, 158, 170, 173
Boulding, K. 114
Brandt Commission 36

Braudel, F. 139
Brown, S. 97
Brundtland Report 14, 35, 37, 102
Brzezinski, Z. 92
Bull, H. 81, 83, 89, 98, 99, 144

Cardoso, F.H. 143
Carr, E.H. 6, 80–1, 88, 94, 98, 144
Carson, R. 34–5
Chipko Movement 71
Clark-Sohn Plan 113, 131, 181n13
Club of Rome, *see* Meadows Report
Cobden, R. 111, 116–17, 126
Cohn, C. 149
Cold War 4
collective action problems 32
Commoner, B. 34
complex interdependence 6–7, 9, 77, 108, 118, 135, 157
Conca, K. 74
Confucius 13
conservationism 14, 61, 166, 171
Convention on International Trade in Endangered Species 40
cosmopolitanism 10
Cox, R. 6, 13, 18, 98, 139, 144–5, 150–1, 154

Daly, H. 35
Darwin, C. 55, 92
Debt-for-nature swaps 38–9
deep ecology 15, 16, 48, 51, 52, 60–2, 65

democracy, direct 16, 64, 173; representative 65
Democritus 23, 98
dependency theory 11, 16, 139, 142–3, 150–1, 154
Der Derian, J. 147
Descartes, R. 28, 43, 98
desertification 3
Deudney, D. 75, 86
Deutsch, K. 113, 121, 123
Devall, B. 62–3
diversity, and political theory 115
Dobson, A. 158
Doyle, M. 78, 99, 107, 109

Earth First! 48, 62
Eckersley, R. 68, 158, 173
ecoanarchism 16, 51, 62, 69 (see also social ecology)
ecocentrism 52
ecofascism 15, 46–7, 50, 102, 172
ecofeminism 17, 51, 69–71, 158
ecosocialism 16, 51, 53, 67–9, 133, 169
education 42, 132–3
Ehrlich, P. 34
Einstein, A. 161
Elshtain, J. 149
enabling state 68, 173
Endangered Species Act (U.S.) 39
Enlightenment, 27, 57, 65, 118, 147, 155, 162
Enloe, C. 149
epistemic communities 11, 113, 129, 135, 137, 146
Etzioni, A. 122
European Union 112, 179n2

Faletto, E. 143
Falk, R. 98, 144, 152
feminism 12–13, 52, 139, 148–9, 155 (see also ecofeminism)
Foreman, D. 52
Foucault, M. 60, 90, 145
Fox, W. 61, 71
Frankfurt School 12, 63, 138, 144, 145

Friends of the Earth 37
functionalism 82–3, 112, 114, 127, 130, 133, 144

Gaia 15, 47–9, 51, 60, 66, 68, 172
Galileo 43
Galtung, J. 152
Game theory 99
Gandhi, M.K. 54, 55, 61, 152
geopolitics 75–6, 86, 103, 156
Giddens, A. 141, 143, 145, 150, 154
Gilbert, F. 79
Gilpin, R. 77, 82, 96–7, 141
global warming 3
globalization 10, 12, 13, 107, 125, 136, 137, 139, 140, 144, 154, 161, 165
Godwin, W. 55
Goldsmith, E. 41
Gramsci, A. 144–5
Green Leviathan 17, 24, 42–6, 50, 73, 103, 135, 137, 171
Green Party, German 69
green revolution 153, 180n12
Greenpeace 37
Grieco, J. 77
Grotius 11, 120, 131
Guha, R. 56
Gurtov, M. 132, 152

Haas, E. 3, 82, 113, 122, 129–30
Haas, P. 11, 129
Haeckl, E. 24
Halliday, F. 135
Hardin, G. 14, 44–5, 166
Hayward, T. 65
hegemonic stability theory 141
hegemony 175n4
Herz, J. 82, 84, 88, 91–2, 95, 99
Hessing, M. 70
Hobbes, T. 7, 9, 43–6, 49, 65, 73, 75, 78–9, 85, 98, 109, 171
Hoffmann, S. 81–2
Holsti, K.J. 83
Homer-Dixon, T. 77, 103
human nature 8, 9

imperialism 1, 58, 86
indigenous rights 173
institutionalism 7
integration theory 113, 121, 123, 140
intergenerational equity 23
international law 108–9, 111, 131

Jary, D. 143
Johansen, R. 125, 130, 132, 152

Kahler, M. 122
Kant, I. 10, 27, 110–11, 126, 176n11
Kaufman, W. 30
Kautilya 7
Keeley, J. 146
Kennedy, P. 86
Keohane, R. 7, 9, 77, 82, 113, 121, 122
Kim, S. 152
Kissinger, H. 92
Kothari, R. 128, 132, 133, 152–3
Krasner, S. 82
Kropotkin, P. 54, 55, 64, 67
Kuehls, T. 155

land ethic 61
Laski, H. 120, 127
League of Nations 94, 111, 120
Leopold, A. 39, 61
Leucippus 23
Leviathan, *see* Hobbes
libertarianism 54
"lifeboat ethics" 14, 45
Linklater, A. 149–50
Lipschutz, R. 74
Litfin, K. 146, 155
Little, R. 10, 115
Locke, J. 43–4, 65, 67
Lovelock, J. 47–8

Machiavelli, N. 7, 9, 78–9, 98, 99, 148
Malthus, T. 29, 33–5
managerialism 14, 41
Manes, C. 62
Marcuse, H. 60

Marx, K. 11, 52, 55, 60, 64, 67–8, 73, 90–1, 138, 154, 162
Marxism, *see* Marx
materialism 179n4
Mathews, J.T. 38
Matthew, R. 10, 123
Mazrui, A. 125, 128, 132, 152
Mazzini, G. 80, 111, 115–118, 120, 128, 129, 132, 133
McKinlay, R.D. 10, 115
Meadows Report 31, 35, 39
mechanicism 9, 72
mercantilism 80
Merchant, C. 56, 70
military establishment 84, 102–3, 173
Mill, J.S. 27, 28–30, 32–3, 35, 45, 64, 65
Mische, P. 132, 152
Mitrany, D. 20, 112, 120–1, 127, 129, 135, 137
modernization 10, 142
monoculture 153, 161
Morgenthau, H. 6, 7, 81, 87–8, 92, 99–100, 113
Morrison, R. 47
mysticism 15, 16, 50, 60, 70
Muir, J. 39, 61
Mumford, L. 54, 56–8, 70

Naess, A. 41, 60–2
National Wildlife Federation 39
natural disasters 3
Nazism 15, 46–7, 51
neo-functionalism 11, 112–13, 121, 129
New International Economic Order 36
new social movements 150–1
Newton, I. 3, 23, 161
Niebuhr, R. 81, 88, 91, 94, 96, 98, 99
Nietzsche, F. 47, 89
nihilism 15, 48, 100
non-governmental organizations (NGOs) 124, 140
North-South issues 1
Norton, B. 32, 52

nuclear weapons 1
Nye, J. 7, 9, 77, 86, 122

Olsen, M. 122
Ophuls, W. 45, 46, 157–8, 173
organicism 64, 70
ozone layer depletion 3, 40, 146

peace studies 114, 122, 128, 151–2
Peloponnesian War, History of, see Thucydides
Peluso, N. 50
Pepper, D. 52, 68
Peterson, V.S. 149
Pinchot, G. 32, 61
Pirages, D. 155
Plato 42–4, 70
Plumwood, V. 71
Polanyi, K. 139
polarity 97
positive peace 56, 102, 152
positivism 9, 12–13, 17–18, 20, 81, 83, 85, 136, 138, 141, 143, 147, 166
postmodernism 12, 52, 145–7, 148, 155, 162, 182n11
postpositivism 142, 145, 151, 157, 162
preservationism 15, 32, 60
pre-Socratic philosophy 22–23
Prince, see Machiavelli

rational choice theory 82, 85, 148
rationalism 14
Ravindra, R. 13
reactionary ecology 14
Realpolitik 76, 78, 80, 103
reflectivism 7, 137
regime theory 2, 114, 121, 122, 135, 136–7, 140, 146, 155, 156–7
religion, and the environment 13, 61
Renaissance 98
Ricardo, D. 67
Richardson, L. 114
romanticism 14, 15
Rosenau, J. 113, 119, 122, 123–5, 127–8, 130, 141

Rousseau, J.J. 45, 64, 65, 78, 109
Russett, B. 128

Sale, K. 66, 158
Sayers, D. 58
Schumacher, E.F. 54, 56, 58–9, 66
self-interest 9
Sessions, G. 62–3
Shiva, V. 56, 71, 153
Sierra Club 39, 61
Simon, J. 31
Smith, A. 27, 65
social contract 43, 45, 161
social Darwinism, 100
social ecology 16, 51, 52, 62, 63–6
soil erosion 3
Sophists 78
Sprout, H. and M. 155, 156
state of nature 8, 88
steady-state 29, 58
stewardship 23
structurationism 141–4, 154, 182n10
sustainable development 1, 3, 36, 37, 41
systems theory 2, 176n1

Taoism 52
technocracy 11, 130
teleology 23, 68, 169
Thoreau, H.D. 2, 54, 55, 61, 155
Thucydides 7, 78, 96, 98
Tickner, J.A. 148
Tocqueville, A. 29
Toynbee, A. 46
trade 120, 126, 133–4, 173
tragedy of the commons 44
transnationalism 122, 135
Treitschke, H. 80, 89–90, 91, 93–4

United Nations Conference on Environment and Development (UNCED) 1, 39
United Nations Conference on the Human Environment (UNCHE) 39
United Nations Environment Programme 39

utilitarianism 3, 14, 19, 20, 24, 26–31, 135

vegetarianism 46
Vernadsky, V.I. 61

Walker, R.B.J. 86, 148
Waltz, K. 6, 8, 87, 95
Wapner, P. 41
war 109–10, 118, 126
Weber, M. 60
Weber, S. 142
Weiss, E.B. 24
Wendt, A. 142
Wight, M. 81
Wilson, W. 111–12, 118, 120, 121, 127
Wolfers, A. 82

Woodcock, G. 64
World Bank (IBRD) 38
World Commission on Environment and Development (WCED), *see* Brundtland Report
World Order Models Project (WOMP) 12, 13, 21, 113–14, 125, 128, 139, 151
World Resources Institute 38
world systems theory 11
World Wildlife Fund 37
Worldwatch Institute 31, 38

Young, O. 11, 129

Zacher, M. 10, 123
Zimmerman, M. 52

International Relations Theory and Ecological Thought

Environmental problems are often international in scope, and a substantial body of academic work has evolved as a result, but this is the first book to bring these intrinsically related fields – international relations theory and ecophilosophy – together.

This book asks what the ecological crisis can teach international relations theorists. It examines the ecological attributes of realism, liberalism, and critical IR theory. Eric Laferrière and Peter Stoett suggest that a critical ecological perspective can make an innovative, and much needed, contribution to IR theory. They also explore the possibility that the international nature of contemporary environmental problems will affect ecological thought in turn.

This groundbreaking book will be a point of departure for all international relations and political theorists, as well as those involved with environmental policy and philosophy.

Eric Laferrière teaches Humanities at John Abbott College, Ste-Anne-de-Bellevue, Quebec. **Peter J. Stoett** teaches International Relations at Concordia University, Montreal.

Environment Politics
Edited by Michael Waller, *University of Keele*,
and Stephen Young, *University of Manchester*

The fate of the planet is an issue of major concern to governments throughout the world and is likely to retain its hold on the agenda of national administrations due to international pressures. As an object of academic study, environmental politics is developing an increasingly high profile: there is a great need for definition of the field and for a more comprehensive coverage of its concerns. This new series will provide this definition and coverage, presenting books in three broad categories:

- new social movements and green parties;
- the making and implementation of environmental policy;
- green ideas.

Titles include:

Global Warming and Global Politics
Matthew Paterson

Politics and the Environment: From Theory to Practice
James Connelly and Graham Smith

International Relations Theory and Ecological Thought: Towards a Synthesis
Eric Laferrière and Peter J. Stoett